Effects of Nuclear Earth-Penetrator and Other Weapons

Committee on the Effects of Nuclear Earth-Penetrator and Other Weapons
Division on Engineering and Physical Sciences

NATIONAL RESEARCH COUNCIL
OF THE NATIONAL ACADEMIES

THE NATIONAL ACADEMIES PRESS
Washington, D.C.
www.nap.edu

THE NATIONAL ACADEMIES PRESS **500 Fifth Street, N.W.** **Washington, DC 20001**

NOTICE: The project that is the subject of this report was approved by the Governing Board of the National Research Council, whose members are drawn from the councils of the National Academy of Sciences, the National Academy of Engineering, and the Institute of Medicine. The members of the committee responsible for the report were chosen for their special competences and with regard for appropriate balance.

This study was supported by funding from the Office of the Under Secretary of Defense (Acquisition, Technology, and Logistics), the National Nuclear Security Administration, and the Defense Threat Reduction Agency under Contract No. HDTRA-04-C-0004 between the National Academy of Sciences and the Defense Threat Reduction Agency. Any opinions, findings, conclusions, or recommendations expressed in this publication are those of the author(s) and do not necessarily reflect the views of the organizations or agencies that provided support for the project.

International Standard Book Number 0-309-09673-1

Copies of this report are available from:

Division on Engineering and Physical Sciences
The Keck Center of the National Academies
500 Fifth Street, N.W.
Washington, DC 20001

Additional copies of this report are available from the National Academies Press, 500 Fifth Street, N.W., Lockbox 285, Washington, DC 20055; (800) 624-6242 or (202) 334-3313 (in the Washington metropolitan area); Internet, http://www.nap.edu.

Printed in the United States of America

THE NATIONAL ACADEMIES

Advisers to the Nation on Science, Engineering, and Medicine

The **National Academy of Sciences** is a private, nonprofit, self-perpetuating society of distinguished scholars engaged in scientific and engineering research, dedicated to the furtherance of science and technology and to their use for the general welfare. Upon the authority of the charter granted to it by the Congress in 1863, the Academy has a mandate that requires it to advise the federal government on scientific and technical matters. Dr. Ralph J. Cicerone is president of the National Academy of Sciences.

The **National Academy of Engineering** was established in 1964, under the charter of the National Academy of Sciences, as a parallel organization of outstanding engineers. It is autonomous in its administration and in the selection of its members, sharing with the National Academy of Sciences the responsibility for advising the federal government. The National Academy of Engineering also sponsors engineering programs aimed at meeting national needs, encourages education and research, and recognizes the superior achievements of engineers. Dr. Wm. A. Wulf is president of the National Academy of Engineering.

The **Institute of Medicine** was established in 1970 by the National Academy of Sciences to secure the services of eminent members of appropriate professions in the examination of policy matters pertaining to the health of the public. The Institute acts under the responsibility given to the National Academy of Sciences by its congressional charter to be an adviser to the federal government and, upon its own initiative, to identify issues of medical care, research, and education. Dr. Harvey V. Fineberg is president of the Institute of Medicine.

The **National Research Council** was organized by the National Academy of Sciences in 1916 to associate the broad community of science and technology with the Academy's purposes of furthering knowledge and advising the federal government. Functioning in accordance with general policies determined by the Academy, the Council has become the principal operating agency of both the National Academy of Sciences and the National Academy of Engineering in providing services to the government, the public, and the scientific and engineering communities. The Council is administered jointly by both Academies and the Institute of Medicine. Dr. Ralph J. Cicerone and Dr. Wm. A. Wulf are chair and vice chair, respectively, of the National Research Council.

www.national-academies.org

COMMITTEE ON THE EFFECTS OF NUCLEAR EARTH-PENETRATOR AND OTHER WEAPONS

JOHN F. AHEARNE, Sigma Xi, The Scientific Research Society, *Chair*
LYNN R. ANSPAUGH, University of Utah School of Medicine
RODNEY C. EWING, University of Michigan
STEVEN A. FETTER, University of Maryland
RICHARD L. GARWIN, IBM Thomas J. Watson Research Center, Fellow Emeritus
SYDELL P. GOLD, Science Applications International Corporation
EUGENE G. GREWIS, Tijeras, New Mexico
THEODORE M. HARDEBECK, Science Applications International Corporation
RAYMOND JEANLOZ, University of California at Berkeley
WILLIAM J. PATTERSON, Albuquerque, New Mexico
GLORIA S. PATTON, Lake Oswego, Oregon
HEINZ W. SCHMITT, Albuquerque, New Mexico
EUGENE SEVIN, Lyndhurst, Ohio
C. BRUCE TARTER, Lawrence Livermore National Laboratory, Director Emeritus
ROBERT H. WERTHEIM, San Diego, California

Staff

JAMES E. KILLIAN, Study Director, National Materials Advisory Board
MARY G. GORDON, Information Officer, Naval Studies Board
SUSAN G. CAMPBELL, Administrative Coordinator, Naval Studies Board
IAN M. CAMERON, Research Associate, Naval Studies Board
SIDNEY G. REED, JR., Consultant, Naval Studies Board

Preface

Section 1033 of the Bob Stump National Defense Authorization Act for Fiscal Year 2003 (Public Law 107-314) directed the Secretary of Defense to request that the National Research Council study the anticipated health and environmental effects of nuclear earth-penetrator and other weapons. Upon request from the Department of Defense, the National Research Council established the Committee on the Effects of Nuclear Earth-Penetrator and Other Weapons (see Appendix A). This report is the product of that committee of technical experts. Some of the material used for this study is classified, and the committee has produced an unclassified report plus a separate classified annex.

Nuclear earth-penetrator weapons, such as those described as robust nuclear earth penetrator (RNEP) weapons, are controversial. Part of the controversy has been about whether a nuclear earth-penetrator could be designed to defeat a deeply buried hard target but not cause any collateral surface damage, such as that attributable to fallout. Another question is whether chemical and biological agents can be defeated by conventional weapons. The committee developed its report in what it believes is a logical approach to address these issues.

To gather information for this report, the committee received briefings from the Department of Defense (DOD), the Department of Energy (DOE), congressional staff, nongovernmental organizations, and individuals, in classified and open sessions. This input included information on potential targets, such as their numbers, location, functions, hardening characteristics, and contents. The committee also received information on after-action investigations in Afghanistan and Iraq. The committee requested that the Defense Threat Reduction Agency (DTRA) and the Lawrence Livermore National Laboratory (LLNL) estimate the numbers of casualties for a range of nuclear weapons' yields and depths of burst for several target areas. Calculations also were done in order to generate lethality contours to estimate numbers of casualties resulting from attacks with non-nuclear weapons on facilities producing or storing weapons of mass destruction.

The committee thanks the many briefers who presented information essential to the writing of this report. They are listed in the agendas provided in Appendix B. Donald Linger of DTRA was consistently helpful, and his deep knowledge greatly assisted our work. Todd Hann, Andrew Grose, Brian Hall, and Michael Phillips of DTRA and Theodore Harvey and Frank Serduke of the LLNL devoted substantial effort to produce the calculations that the committee needed. The committee also thanks Milton Minneman of the DOD, our project liaison, who assisted in obtaining reports needed for the study.

Finally, the committee thanks the dedicated staff of the National Research Council: Dixie Gordon, who resolved the many difficulties associated with handling classified material; Susan Campbell, who capably saw to the logistics for our many meetings over a short period of time and helped with report production; Ian Cameron, who supported the conduct of the meetings and also the production of the report; and, in particular, our study director, James Killian, who kept the study on track and provided invaluable support to the committee members and to me.

John F. Ahearne, *Chair*
Committee on the Effects of Nuclear Earth-Penetrator and Other Weapons

Acknowledgment of Reviewers

This report has been reviewed in draft form by individuals chosen for their diverse perspectives and technical expertise, in accordance with procedures approved by the National Research Council's Report Review Committee. The purpose of this independent review is to provide candid and critical comments that will assist the institution in making its published report as sound as possible and to ensure that the report meets institutional standards for objectivity, evidence, and responsiveness to the study charge. The review comments and draft manuscript remain confidential to protect the integrity of the deliberative process. We wish to thank the following individuals for their review of this report:

James L. Bacon, Pine Bluff, Arkansas,
Sidney D. Drell, Stanford University,
Daniel J. Fink, Potomac, Maryland,
David R. Franz, Midwest Research Institute,
Thomas A. Griffy, University of Texas at Austin,
Michael M. May, Stanford University,
Fred Mettler, Jr., University of New Mexico,
Richard Mies, Science Applications International Corporation,
Bruce Napier, Pacific Northwest National Laboratory, and
George M. Whitesides, Harvard University.

Although the reviewers listed above provided many constructive comments and suggestions, they were not asked to endorse the conclusions or recommendations, nor did they see the final draft of the report before its release. The review of this report was overseen by Hyla S. Napadensky, Napadensky Energetics, Inc. (retired), and Chris G. Whipple, ENVIRON International Corporation. Appointed by the National Research Council, they were responsible for making certain that an independent examination of this report was carried out in accordance with institutional procedures and that all review comments were carefully considered. Responsibility for the final content of this report rests entirely with the authoring committee and the institution.

Contents

APPENDIXES

Summary

The tasking for the Committee on the Effects of Nuclear Earth-Penetrator and Other Weapons, stated in Section 1033 of the Bob Stump National Defense Authorization Act for Fiscal Year 2003 (Public Law 107-314), is included in Chapter 1 of this report. The charge requests an examination of the anticipated health and environmental effects of (1) a nuclear earth-penetrator weapon that would enhance ground-shock coupling to destroy deep underground or other hard targets, (2) a nonpenetrating nuclear weapon that would also be used against deeply buried or other hard targets, and (3) conventional weapons used against facilities for the storage or production of weapons of mass destruction. Study of the effects on civilian populations and on U.S. military personnel who carry out operations or battle damage assessment in the target area is specified.

To provide a more complete analysis of the issues, the committee expanded its study to consider the effects of nuclear weapons used against facilities for the storage of chemical or biological agents. It also considered the effects of nuclear bursts that can be described as locally fallout-free, because the weapon is detonated well above the ground surface.

The committee received many briefings from a wide variety of public and government sources and reviewed classified reports from the Department of Defense (DOD) and the Department of Energy (DOE). Although this report is unclassified, the committee also produced a separate classified annex, which does not modify any of the study's conclusions but provides supporting material.

CONCLUSIONS

The committee's major conclusions are the following:

Conclusion 1. Many of the more important strategic hard and deeply buried targets are beyond the reach of conventional explosive penetrating weapons and can be held at risk of destruction only with nuclear weapons. Many—but not all—known and/or identified hard and deeply buried targets can be held at risk of destruction by one or a few nuclear weapons.

Conclusion 2. Nuclear earth-penetrator weapons (EPWs) with a depth of penetration of 3 meters capture most of the advantage associated with the coupling of ground shock. While additional depth of penetration increases ground-shock coupling, it also increases the uncertainty of EPW survival. To hold at risk hard and deeply buried targets, the nuclear yield must be increased with increasing depth of the target. The calculated limit for holding hard and deeply buried targets at risk of destruction with high probability using a nuclear EPW is approximately 200 meters for a 300 kiloton weapon and 300 meters for a 1 megaton weapon.

Conclusion 3. Current experience and empirical predictions indicate that earth-penetrator weapons *cannot* penetrate to depths required for total containment of the effects of a nuclear explosion.

Conclusion 4. For the same yield and weather conditions, the number of casualties from an earth-penetrator weapon detonated at a few meters depth is, for all practical purposes, equal to that from a surface burst of the same weapon yield. Any reduction in casualties due to the use of an EPW is attributable primarily to the reduction in yield made possible by the greater ground shock produced by buried bursts.

Conclusion 5. The yield required of a nuclear weapon to destroy a hard and deeply buried target is reduced by a factor of 15 to 25 by enhanced ground-shock coupling if the weapon is detonated a few meters below the surface.

Conclusion 6. For attacks near or in densely populated urban areas using nuclear earth-penetrator weapons on hard and deeply buried targets (HDBTs), the number of casualties can range from thousands to more than a million, depending primarily on weapon yield. For attacks on HDBTs in remote, lightly populated areas, casualties can range from as few as hundreds at low weapon yields to hundreds of thousands at high yields and with unfavorable winds.

Conclusion 7. For urban targets, civilian casualties from a nuclear earth-penetrator weapon are reduced by a factor of 2 to 10 compared with those from a surface burst having 25 times the yield.

Conclusion 8. In an attack on a chemical or biological weapons facility, the explosive power of conventional weapons is not likely to be effective in destroying the agent. However, the BLU-118B thermobaric bomb, if detonated within the chamber, may be able to destroy the agent. An attack by a nuclear weapon would be effective in destroying the agent only if detonated in the chamber where agents are stored.

Conclusion 9. In an attack with a nuclear weapon on a chemical weapons facility, civilian deaths from the effects of the nuclear weapon itself are likely to be much greater than civilian deaths from dispersal of the chemical agents. In contrast, if the target is a biological weapons facility, release of as little as 0.1 kilogram of anthrax spores will result in a calculated number of fatalities that is comparable on average to the number calculated for a 3 kiloton nuclear earth-penetrator weapon.

Additional conclusions are presented in Chapter 9 of this report.

The committee notes that although some scenarios show substantial nuclear-radiation-induced fatalities, military operational guidance is to attack targets in ways to minimize collateral effects. Calculated numbers of fatalities to be expected from an attack on an HDBT might be reduced by operational planning and employment tactics. Assuming that other strategic considerations permit, the operational commander could warn of a nuclear attack on an HDBT or could time such an attack to take advantage of wind conditions that would reduce expected casualties from acute and latent effects of fallout by factors of up to 100, assuming that the wind conditions were known well enough and were stable and that defenses against the attack could not be mobilized. However, a nuclear weapon burst in a densely populated urban environment will always result in a large number of casualties.

BACKGROUND

Potential U.S. adversaries worldwide are using underground facilities to conceal and protect leaders, military and industrial personnel, weapons, equipment, and various other assets and activities. These facilities include hardened surface bunkers and tunnel facilities deep underground. Specifically, many underground command, control, and communications (C^3) complexes and missile tunnels are between 100 and 400 meters below the surface, with the majority less than 250 meters deep. A few are as deep as 500 or even 700 meters, in competent granite or limestone rock.

The activities in such underground facilities pose a potentially serious threat to U.S. national security. As a generic term, "hard and deeply buried targets" refers to *all* types of intentionally hardened targets, *either aboveground or belowground.* The DOD estimates that 10,000 HDBTs exist in the territory of potential adversaries worldwide. Of the estimated 10,000 HDBTs, about 20 percent have a major strategic function; of that 20 percent, about half are near or in urban areas. More than 100 HDBTs could be candidates for targeting with a robust nuclear earth penetrator (RNEP) weapon, if one were developed (the Robust Nuclear Earth Penetrator program is currently an engineering feasibility study). (In Chapter 3, see the section entitled "Current Robust Nuclear Earth Penetrator Program.")

Although much of the congressional discussion in this area has been about the RNEP weapon, a more general term is "earth-penetrator weapon." The EPW is designed to detonate below the ground surface after surviving the extremely high shock and structural loading environments that result during impact and penetration.

The DOE national laboratories and DOD laboratories have maintained EPW programs and testing activities since the 1960s, resulting in more than 1,000 representative non-nuclear penetration tests that are recorded in the Sandia National Laboratories Earth Penetration Database. Penetration tests have been conducted at various impact angles, angles of attack, and velocities into undisturbed geologic targets to provide insight into how the physical properties of a penetrator affect its ability to penetrate.

The greatest uncertainty in predicting EPW depth of penetration and structural survival of the weapon until detonation arises from the inherently heterogeneous nature of the local subsurface geology. This uncertainty can be countered to some degree by designing an EPW to be as rugged as possible, consistent with mission and system requirements. Calculated penetration depths depend on the mechanical properties of the earth materials at the target point. For example, for the same penetrator and velocity, calculations give penetration depths of 100 meters in a silty clay soil, 30 meters in low-strength rock, and 12 meters in medium-strength rock; and the maximum depth in soil can vary by ±20 percent. Deeper EPW penetration is generally better for target destruction because the ground-shock coupling increases with deeper depth of burst (DOB), although most of the advantage is obtained in the first few meters.

The current nuclear EPW is the B61-11, which uses the B61-7 nuclear weapon components and was developed to replace the B53 gravity bomb. The Robust Nuclear Earth Penetrator program is an engineering feasibility study to determine if it is possible to design an earth-penetrator weapon system that uses the major components of an existing weapon system and can hold at risk of destruction a significantly larger number of HDBTs than could the B61-11.

NUCLEAR EARTH-PENETRATOR WEAPON

Although conventional high-explosive weapons can penetrate at least as deep as a nuclear EPW can, if not deeper, conventional weapons are not likely to be effective against targets that the penetrator cannot reach. For destroying targets near the surface, however, either nuclear or conventional weapons may be effective. Because of the radiation doses and much higher temperatures associated with their detonation, nuclear weapons are expected to be more effective than conventional weapons at destroying biological or chemical agents.

The major advantage of an EPW over a surface or aboveground burst is the effectiveness with which energy is transmitted into the ground. The ground-shock-coupled energy of an earth-penetrator weapon approaches 50 percent with increasing depth of burst, and is effectively fully coupled at a scaled DOB of about 2.3 $m/Y^{1/3}$ (where m is depth of burst in meters and Y is yield in kilotons).[1] The ground-shock-coupling factor has already risen to 15 to 25 for a 300 kiloton EPW at 3 meters' depth of burst (scaled DOB of about 0.5 $m/Y^{1/3}$). Calculations indicate that such a weapon is capable of severely damaging tunnels in a competent granite site down to depths of around 150 meters with a 0.95 probability. A nonpenetrating nuclear weapon capable of causing the same damage would have a yield of about 6 megatons. To be fully contained (i.e., with no venting of radioactive gases), a 300 kiloton weapon would have to be detonated at the bottom of a carefully stemmed emplacement hole about 800 meters deep. Because the practical penetration depth for an EPW is a few meters—a small fraction of the depth for full containment—there will be blast, thermal, initial nuclear radiation, and fallout effects from use of an EPW.

The effectiveness of nuclear weapons against deeply buried targets can be estimated by calculating the intensity of the ground shock in the vicinity of the buried target in relation to the hardness of the target. There is a reasonably extensive experimental database, *Effects Manual Number 1 (EM-1)*,[2] covering the various physics regimes governing the energy-coupling process. Uncertainties associated with estimates of energy coupling into the ground are far greater for near-surface airbursts than for buried bursts, and they depend on how well the actual burst location and details of weapon energy output are known.

The Defense Threat Reduction Agency and DOE have invested considerable resources to develop computational methods for predicting the ground-shock environments at depth from both high-explosive and nuclear bursts. This is a complicated problem owing to various shock-attenuation mechanisms—such as inelastic effects, hysteresis, fracture, and dilatation—and geometric effects due to divergence of the stress waves and the presence of layers, interfaces, faults, and joints throughout the target area. The directly applicable U.S. experimental database, EM-1, is limited to the results of data on eight underground nuclear tests in which tunnels of various construction types were exposed to damaging ground-shock levels of nuclear bursts in a few types of rock geologies. Only two of these tests were dedicated to experiments on engineered structures in competent granite geology. The others were add-on experiments to underground nuclear tests conducted for different purposes on engineered structures in relatively soft tuff geology.

Calculations show that both surface-burst and earth-penetrating nuclear weapons must be delivered with high accuracy in order to have a high probability of destroying hard and deeply buried targets. For example, a circular error probable (CEP) of less than 60 meters is needed for a 1 megaton contact burst for targets of at most 125 meters' depth to be held at risk with a 0.95 probability of severe damage. For an EPW, a yield of 300 kilotons eases the accuracy requirements to a CEP of 110 meters or less, with targets potentially as deep as 225 meters held at risk with a 0.95 probability of severe damage.

COLLATERAL EFFECTS

The primary goal for any nuclear weapon is the deterrence of a potential adversary by the ability to hold the adversary's most-valued assets at risk of destruction. To contribute to deterrence, the weapon should be capable of defeating those assets. The use of a weapon to accomplish the goal of target defeat or destruction will have accompanying collateral effects that, in the case of nuclear weapons, can be extremely large.

Modeling collateral effects is a multistage process. Estimated first are deaths and serious injuries due to "prompt" (i.e., occurring immediately after detonation) effects—air blast, thermal effects, and initial nuclear radiation. Second, the downwind transport and deposition of radioactive material produced by the explosion are modeled. Third, the dose from external radiation from ground-deposited fallout is calculated. Fourth, the health effects of exposure to radiation are estimated in those populations that survive the prompt effects of the explosion.

Two computer programs are in wide use to model collateral effects. The Hazard Prediction and Assessment Capability (HPAC) code was developed by the Defense Threat Reduction Agency and its predecessor agencies to analyze nuclear, chemical, and biological releases for military studies and operational planning. The K-Division Defense Nuclear Agency Fallout Code (KDFOC) was developed by Lawrence Livermore National Laboratory to model fallout from nonweapon Plowshare tests, which involved nuclear explosives designed to produce craters with minimal fallout. Both computer codes are calibrated to available data from nuclear tests conducted at the Nevada Test Site. They differ somewhat, including their treatment of prompt casualties due to blast and radiation, wind transport, and the prediction of casualties associated with a given level of radiation from fallout.

Fallout is a long-studied and experimentally measured feature of many nuclear weapons tests. When a nuclear weapon is exploded underground, a sphere of extremely hot, high-pressure gases is formed, which includes vaporized weapon residues and ground materials, that is the equivalent of the fireball in an airburst or surface burst. If the subsurface burst is at a shallow depth, the pressure of the explosion, uncompensated by similar pressure above the surface, will throw rock, soil, and weapon material into the air.

Fallout is determined primarily by the fission yield of the weapon, the amount and constitution (hence activation) of entrained mass, the injection height distribution, the particle size distribution, and subsequent atmospheric transport. Surface geology is critical. The prediction of fallout for shallow buried bursts is uncertain because the United States has performed only three tests at depths shallower than 20 scaled meters, and none of these tests was in rock. Another feature of a buried or surface burst is the base surge. The base surge begins to form as the growth of the crater stops and entrained material in the column begins to fall and expand radially along the ground surface. For depths of burst of 2 to 3 scaled meters, the fraction of activity in the base surge is typically less than a few percent of the total activity.

Presumably, nuclear EPWs would not be used for surface and near-surface point targets, especially if other options were available that were effective and could ameliorate the collateral damage due to fallout. Calculations have been done for the so-called fallout-free height of burst (HOB). The fallout-free HOB, as its name implies, is sufficiently high that the fireball produced by the nuclear explosion does not touch the ground surface. In the absence of rain, the explosion therefore is not expected to generate significant local fallout, because no surface material is activated, entrained, lofted, or dispersed, and the weapon residues are present in the form of fine particles that will remain airborne for weeks or years. For a 1 megaton weapon the fallout-free HOB is about 900 meters. The nuclear weapons at both Hiroshima and Nagasaki were detonated above the fallout-free HOB and produced no significant local fallout.

Thermal radiation from the fireball may make fire a significant collateral effect, especially for airburst and surface-burst nuclear weapons. The potential for fire damage depends on the nature of the burst and the surroundings. Fires can be an indirect effect of destruction caused by a blast wave, which can upset stoves, furnaces, gas lines, and so on.

The committee asked Lawrence Livermore National Laboratory (LLNL) and the Defense Threat Reduction Agency (DTRA) to run several scenarios involving three typical targets, a range of yields, and both surface and EPW-depth bursts. Once differences in input variables are removed, the LLNL and DTRA results are comparable within the uncertainties in the estimated parameters. The results of these calculations for several scenarios and weapons yields are presented in Chapter 6 and form the basis for several of the committee's conclusions. In addition to the conclusions stated above, significant results include the following:

1. Any reduction in the number of casualties owing to the use of a nuclear earth-penetrator weapon compared with the number of casualties from a surface burst is due primarily to the reduction in yield by a factor of about 25 that is made possible by the greater coupling of the released energy to the ground shock for a buried detonation.
2. For rural targets, the use of a nuclear earth-penetrator weapon is estimated to reduce casualties by a factor of 10 to 100 relative to a nuclear surface burst of equivalent probability of damage.
3. Wind patterns can have an enormous effect on the number of casualties resulting from fallout. For targets in large urban centers, fatalities from acute and latent effects from fallout can vary by more than a factor of 10. For targets outside cities, fatalities from fallout can vary by more than a factor of 100, depending on population distribution and wind direction.
4. The estimated number of deaths and injuries resulting from a nuclear attack depends on many variables, including weapon yield and design, depth of burst, weather conditions, and population distribution and sheltering during and after the attack. The estimated number of casualties ranges over four orders of magnitude—from hundreds to over a million—depending on the combination of assumptions used.

The committee advises readers to keep in mind that the foregoing are the results of model calculations and that they have significant uncertainty due to uncertainties in the physical model inputs (e.g., the definition of the source term), boundary conditions (e.g., weather conditions and population distribution), and the paucity of relevant experience against which the modeled results can be validated. For each of the model calculations, a range of boundary conditions has been assumed. Uncertainty inevitably exists in such calculations, and the scale of these uncertainties is essential to understanding the results of the calculations and the findings of this committee. The uncertainties are of three types: scenario uncertainty, data uncertainty, and conceptual model uncertainty.

CHEMICAL AND BIOLOGICAL AGENTS

The committee's task included examination of the use of conventional weapons against facilities for the storage or production of weapons of mass destruction. The committee addressed the ability of conventional and nuclear earth-penetrator weapons to effectively destroy buried production facilities, stores, and weapons containing chemical agents and biological agents. The Department of Defense Global Strike Mission requires the capability to deliver rapid, extended-range, precision kinetic (nuclear and conventional) and nonkinetic weapons in support of theater and national objectives. Many conventional high-explosive weapons are currently available and under development to support this mission.

Sufficient knowledge from intelligence assessments is of paramount importance to both weapon choice and targeting. Some of the key elements for selecting the weapon type and its impact point(s) include knowing the placement of the storage containers for chemical or biological agents; knowing whether the agents are in production or, if already produced, the type of storage containers and the material of which they are constructed; and knowing the amount of agent in containers. This information is in addition to knowing the depth of the facility and its structure. If agents are in deeply buried facilities that are crushed or rendered unusable, and the fracture zone created by the explosion does not open a channel between the facility and the surface, the probability of agent ejection after impact and detonation is very low.

Biological and chemical agents degrade after they are released into the air. The atmospheric degradation of agents occurs as a result of several mechanisms, such as photochemical reactions, exposure to radiant energy, and atmospheric chemistry. Biological agents are especially susceptible to atmospheric degradation, as their viability decreases depending on levels of radiant solar energy, oxygen, relative humidity, temperature, ozone, and hydroxyl radicals. Chemical agents decompose mainly owing to photochemical processes in the atmosphere, such as reactions with ozone, hydroxyls, and industrial pollutants. Both decay and decomposition are more pronounced during the daytime owing to ultraviolet (UV) radiation and the increase in reactivity with atmospheric components. Therefore, the amount of exposure to solar energy generally tends to determine the rates of degradation. Also, a key factor in the loss of viability or toxicity is the length of exposure to these atmospheric elements and conditions.

Even if large amounts of chemical agents were released, substantial lethal areas would result only under very stable meteorological conditions. The agents differ in how they disperse, but exposure to rain and sunlight reduces their effectiveness. In the case of biological agents, only spores are relatively immune to destruction by UV rays.

Therefore, for a daytime attack, biological agents such as smallpox and tularemia are of relatively low danger except in the immediate vicinity of the explosion, and then for only a short period. Anthrax spores and those of other disease agents are more UV-resistant and can withstand high temperatures. Nevertheless, the data that the Centers for Disease Control and Prevention gathered from the anthrax experience a few years ago and from areas in the United States where anthrax is endemic indicate that few cases of the disease occur from wide exposure to spores after they have entered the ground. Not all chemical agents (VX, mustard, lewisite) aerosolize. They also are similar to anthrax spores in being unaffected by UV. If they are ejected following an explosion, they contaminate the immediately surrounding ground area.

HAZARDS TO U.S. MILITARY PERSONNEL

The committee also addressed the hazards to U.S. military personnel from entering an area after use of an earth-penetrator weapon. Because the committee concluded that such a weapon would produce local fallout, the hazards are similar to those faced by troops entering an area after a surface burst. For equivalent-target damage, because of the substantially smaller EPW yield, the local effects are reduced significantly, but not eliminated.

Current analytical tools have an overall propagated uncertainty no smaller than one order of magnitude (factor of 10), and likely in the range of 10 to 100, for estimates of casualties resulting from a nuclear attack. This conclusion is founded both on evaluation of the underlying calculations (source terms, transport models, grid resolution, and so on) and their experimental validation and on a review of the variability in results that can be obtained for different scenarios when considering plausible ranges in parameters.

At least three key sensitivities affect estimates of military effectiveness and casualties associated with use of a nuclear EPW or a nuclear surface-burst weapon:

1. Target location, especially urban versus rural;
2. Accuracy of weapons delivery (circular error probable) and precise knowledge of target location and structure, as military effectiveness depends strongly on a combination of accurate delivery and yield; and
3. Estimates of the source, transport, and influence on populations of the effects of a nuclear explosion, as these can be highly variable (by factors of up to about 10 to 1,000, depending on assumptions).

One additional sensitivity affects estimates of the effects of the nuclear EPW:

4. Functionality after penetration, especially as influenced by target heterogeneity and its uncertainty (e.g., local geology or complex structures in urban areas).

NOTES

1. Scaled depth of burst (DOB) is a normalization of the actual depth (or height) of a burst based on weapon yield to that for a 1 kiloton weapon. This is determined by $DOB/Y^{1/3}$. Thus, the scaled DOB and actual DOB are the same for a 1 kiloton EPW. For example, a 1 kiloton weapon buried 3 meters has a 3 meter scaled DOB, whereas a 300 kiloton weapon buried at the same depth of 3 meters couples its energy to the ground as if it were a 1 kiloton weapon buried at an actual depth of about 0.45 meter; that is, $3/300^{1/3} = 3/6.67 = 0.45$.

2. Defense Nuclear Agency. 1991. *Effects Manual Number 1 (EM-1),* Chapter 3, "Cratering, Ejecta and Ground Shock," DNA-EM-1-CH-3, Alexandria, Va., December.

1

Introduction

BACKGROUND

The Department of Defense (DOD) estimates that there are 10,000 known or suspected hard and deeply buried targets (HDBTs) worldwide as identified by the Defense Intelligence Agency. Of that number, about 20 percent have a major strategic function, and of those, about half are in or near urban areas. HDBTs are used for the protection of senior leaders, command and control functions, and storage of weapons of mass destruction (WMD), among other purposes. Some of them are buried in rock at depths greater than 300 meters, and some are hardened to withstand overpressures of about 1 kilobar.

A U.S. military requirement exists for capabilities to hold these HDBTs at risk. Past U.S. capabilities to satisfy this requirement for the deepest known HDBTs relied on a nonpenetrating, air-delivered, nuclear bomb of the largest yield in the inventory, the B53, now retired. Another existing nuclear weapon, the B61-7, was modified to become the B61-11, so as to have limited penetration capabilities.

Current DOD plans are to develop capabilities that can provide several options to hold HDBTs at risk. These include strike operations involving nuclear and non-nuclear weapons, Special Forces operations, and nonkinetic approaches (e.g., information operations). For all targets, the lowest yield would be used to achieve necessary destruction while at the same time minimizing collateral damage. To further reduce reliance on nuclear weapons, the 2002 *Nuclear Posture Review*[1] called for the development of high-precision conventional weapons to replace nuclear systems wherever possible.

The Department of Energy's (DOE's) National Nuclear Security Administration (NNSA) and the Air Force are conducting a 3-year study (now in its second year), including early research and development (R&D), toward a robust nuclear earth penetrator (RNEP) weapon, which is to be based on one of two existing nuclear designs, the B61-7 and B83, each to be studied at one of the two nuclear weapons laboratories. At present, only the B83 part of the study is funded. At the study's conclusion, NNSA will state which (if either) of the two competing approaches it would recommend for further R&D. At present, no plans exist for conducting nuclear tests related to RNEP or other nuclear weapons.

Previous congressional actions related to the RNEP have included limiting the obligation of funds for the RNEP study pending a report from the DOD and DOE "that sets forth: (1) the military requirements for the RNEP; (2) the nuclear weapons employment policy for the RNEP; (3) the detailed categories or types of targets that the RNEP is designed to hold at risk; and (4) an assessment of the ability of conventional weapons to address the same types of targets that the RNEP is designed to hold at risk."[2] This congressional directive does not mention collateral effects, but the responding reports from the DOD and DOE discuss (without details) the general expectation that collateral effects would occur if the RNEP were to be used, and that such effects would be much larger if a nonpenetrating weapon were employed to destroy the same HDBTs.

The Defense Threat Reduction Agency (DTRA) and the Air Force have conducted and are continuing to conduct studies, modeling and simulations, and non-nuclear tests to achieve a better understanding of the complex phenomena involved if nuclear or non-nuclear weapons are exploded in or near the location of hardened WMD storage or production facilities. The Thermobaric Weapon Demonstration is an ongoing program to provide an air-to-ground weapon capability to functionally defeat hard and deeply buried tunnel targets—with an emphasis on those for protecting leadership and command and control facilities. Additionally, DTRA has a conventional Counterforce Agent Defeat program that is working specifically on developing weapons and weapon payloads effective against facilities containing chemical and biological agents. One of its products was the BLU-119/B agent defeat weapon, which uses a 2,000 lb blast/fragmentation warhead (modified MK-84) developed and fielded as the Quick Reaction program for Operation Iraqi Freedom to damage fixed biological and chemical targets without contaminating the area. In the early 1990s, there were studies and some R&D on a low-yield (less than 5 kiloton) precision nuclear weapon for the destruction of chemical and biological agents in hardened facilities. In 1994, under the Defense Authorization Act,[3] Congress prohibited R&D that could lead to the production by the United States of a new low-yield nuclear weapon, including a precision low-yield warhead, which as of the date of the law's enactment had not yet entered into production. This restriction was repealed in 2004 for R&D up to but not including the engineering development phase.

STATEMENT OF TASK

Section 1033 of the Bob Stump National Defense Authorization Act for Fiscal Year 2003 (Public Law 107-314) directed the Secretary of Defense to request that the National Academy of Sciences study the anticipated health and environmental effects of nuclear earth-penetrator and other weapons.

As requested, the study examined the following:

1. The anticipated short-term and long-term effects of the use by the United States of a nuclear earth-penetrator weapon on the target area, including the effects on civilian populations in proximity to the target area at the time of or after such use and the effects on the United States military personnel who after such use carry out operations or battle damage assessments in the target area.
2. The anticipated short-term and long-term effects on civilian populations in proximity to a target area:
   a. if a nonpenetrating nuclear weapon is used to attack a hard and deeply buried target; and
   b. if a conventional high-explosive weapon is used to attack an adversary's facilities for storage or production of weapons of mass destruction and, as a result of such attack, radioactive, nuclear, biological, or chemical weapons materials, agents, or other contaminants are released or spread into populated areas.

The National Research Council, the operating arm of the National Academies, convened the Committee on the Effects of Nuclear Earth-Penetrator and Other Weapons under the auspices of the Division on Engineering and Physical Sciences to conduct the study. The committee received briefings from the DOD, DOE, congressional staff, nongovernmental organizations, and individuals, in both classified and open sessions. To assist in its considerations of Congress's questions, the committee asked the DTRA and Lawrence Livermore National Laboratory to estimate the number of civilian casualties for a range of nuclear weapons yields and depths of burst for possible target areas.

COMMITTEE'S UNDERSTANDING AND ASSUMPTIONS

The issues raised in the statement of task are stated briefly and refer to specific combinations of factors. Here, the committee makes clear its understanding of the terms and factors involved. It assumes that the health and environmental effects of nuclear earth-penetrator and other weapons would be produced by an attack expected to be militarily effective. "Hard and deeply buried target" is taken to indicate primarily hard targets, which may be buried at various depths. "Deeply buried" refers to depths beyond those of craters formed by the largest nuclear weapon that might be used to destroy such targets.

The statement of task refers explicitly to WMD only in the case of a conventional weapon being used to attack facilities containing WMD and does not mention WMD in connection with the effects of a nuclear attack on HDBTs. The committee thinks that a fully effective attack on an HDBT, whether a nuclear or non-nuclear attack, should involve the destruction or complete sealing off of contained WMD.

In general, in this report, discussion of WMD often lumps chemical and biological agents together. Chemical and biological agents, however, have distinct properties that lead to differences in the effects of their release and in the difficulty of their destruction.

In a comparison of the effects of nuclear explosions, consideration must be given to other effects besides radioactivity. Local blast and thermal effects of near-surface nuclear explosives can be more lethal than the radioactivity.

To understand the statement of task's asymmetry between nuclear and conventional weapons, the committee heard from congressional staff involved in developing the task statement and from the DOD sponsors. It was clear that tasks 1 and 2a were the key issues and would require extensive analysis, including substantial original calculations. Task 2b, regarding conventional weapons, was of much lower importance and was to be examined if time and resources permitted. Unfortunately, the efforts on tasks 1 and 2a were more demanding than estimated, particularly because of the need to develop and analyze the material presented in Chapter 6. Consequently, less time and fewer resources were available to work on task 2b.

ORGANIZATION AND CONTENT OF THIS REPORT

With the understandings and assumptions specified above, the committee organized its response to the statement of task as follows.

Chapter 2 defines HDBTs and discusses their types, locations, functions, size, overall numbers, different depths in different geologies, and trends in types of HDBTs. It gives examples of some HDBTs with details and emphasizes the need for accurate target intelligence.

Chapter 3 defines earth-penetrator weapons (EPWs) and discusses the history of EPW technology, key penetrator characteristics, geologies in which experiments have been done, empirical equations for predicting EPW penetration, maximum credible penetration depths, the B61-11 EPW, the current Robust Nuclear Earth Penetrator program, and some characteristics of nuclear weapons for surface burst.

Chapter 4 addresses the calculated effects of nuclear weapons against hard and deeply buried targets, with particular attention to target destruction, tools to calculate damage, and the uncertainties involved.

Chapter 5 discusses collateral effects of the fallout from air, surface, and underground bursts, as well as two computer programs that are in wide use to model the effects of release of hazardous chemical, biological, radiological, and nuclear materials into the atmosphere and their effects on civilian and military populations. The computer programs discussed are the Hazard Prediction and Assessment Capability code (HPAC) developed by DTRA and its predecessor agencies and those used by the Lawrence Livermore National Laboratory—the NUKE code from Sandia National Laboratories to model the prompt effects of a nuclear explosion and the K-Division Defense Nuclear Agency Fallout Code (KDFOC) to analyze the spread of radioactivity. Also discussed are the uncertainties involved in the use of computer modeling and simulation.

Chapter 6 addresses the health and environmental effects of nuclear explosions. Computer modeling and simulation calculations are presented for notional representative targets attacked with weapons over a range of yields. In addition, health effects of attacks on chemical and biological weapons facilities are addressed.

Chapter 7 discusses conventional high-explosive weapons, both current and under development for direct or indirect attack, to support the emerging Global Strike Mission.

Chapter 8 discusses sources of uncertainty—factors to which the results of calculations such as those discussed in Chapter 6 are most sensitive—and compares the variations in effects due to uncertainty with the variations expected from substituting an EPW for a surface burst. It addresses effects of uncertainty regarding target location, geology, accuracy of delivery, ability of a weapon to function, and model inputs, as well as effects of uncertainty in the models themselves and in analytical tools.

Chapter 9 presents the committee's conclusions.

The four appendixes provide supplemental and study-process-related information.

NOTES

1. Department of Defense. 2002. *Nuclear Posture Review (U)*, Washington, D.C. (Classified).
2. National Defense Authorization Act for Fiscal Year 2003 (House Rpt. 107-772, sec. 3146, P.L. 107-314).
3. National Defense Authorization Act for Fiscal Year 1994 (P.L. 103-160, 107 Stat. 1547).

2

Hard and Deeply Buried Targets

Potential U.S. adversaries worldwide are using intentionally hardened facilities to conceal and protect their leaders, military and industrial personnel, weapons, equipment, and other assets and activities. Such facilities, called hard and deeply buried targets (HDBTs), are a serious challenge to U.S. national security objectives of maintaining the capability to hold such adversary assets at risk. Ranging from hardened, surface bunker complexes to tunnel facilities deep underground, HDBTs are typically large, complex, and well concealed, incorporating strong physical security, modern air defenses, protective siting, multifaceted communications, and other important features that make many of them able to survive attack by conventional weapons. Potential adversaries are increasingly locating HDBTs in basements of multistory buildings located in urban settings, complicating attack planning and increasing the risk of serious collateral effects. This situation places a premium on achieving accurate target characterization so as to obtain the required lethality from precisely delivered weapons during a strike.

Many HDBTs are of a shallow "cut and cover" design, with an equivalent concrete structural overburden of less than 3 meters' thickness. This type of facility typically has a tactical function, such as providing support for artillery or missile launchers. Many such facilities can be held at risk by current weapons, or weapons under development if deployed, if the numbers of U.S. weapons are adequate, accurate target location coordinates are known, and adversaries' defenses are overcome. The missile operations tunnels and armament bunkers in some theaters are particularly troublesome because of their sheer numbers, protective berms, and the strategic positioning of their entrances and exits away from direct routes of attack.

Hundreds of much harder facilities (with a concrete overburden equivalent of 20 to 100 meters) protect strategic capabilities (e.g., leadership, command and control, weapons of mass destruction [WMD]) and were built using either conventional drill-and-blast tunneling techniques or more modern mining equipment. These are typically equipped with redundant ventilation, power, and communications systems. U.S. capabilities to place these types of facilities at risk are challenged not only by the depths of burial and redundancies in critical functional systems, but also by sophisticated techniques for camouflage, concealment, and deception, and some collocation of HDBTs in civilian areas. Such facilities, which conceal and protect an adversary's most valued strategic capabilities, are described in more detail in this chapter.

BASIC DEFINITIONS

Hard and Deeply Buried Targets

The generic term "hard and deeply buried targets" refers to all types of intentionally hardened targets, either aboveground or belowground, that are designed to withstand or minimize the effects of kinetic weapons.

Underground Facility

The generic term "underground facility" refers to all types of underground hardened structures and facilities regardless of their depth.

Hardened Structure

The generic term "hardened structure" refers to a structure that is intentionally strengthened to provide protection from the effects of kinetic weapons. All hardened structures can be further grouped into one of three types, defined by the location of the roof of the structure's functional workspace:

- *Aboveground hardened structure.* The roof of the functional workspace is above the ground surface. Included in this category are earth-bermed structures that are not exposed directly to weapons effects and for which air blast effects are reduced owing to the aerodynamic shape of the berm.
- *Shallow underground structure.* The roof of the functional workspace is between the ground surface level and 20 meters deep.
- *Deep underground hardened structure.* The roof of the functional workspace is covered by 20 or more meters of soil and/or rock.

Strategic Hard and Deeply Buried Target

The term "strategic hard and deeply buried target" refers to those HDBTs that perform a strategic function, such as command and control of military forces, protection of national leadership, WMD production or storage, and ballistic missile production, storage, or launch. The proposed nuclear earth-penetrator weapon (EPW) is being designed to defeat this target class.

BASIC FACTS AND ESTIMATES

Following is a concise list of background facts and estimates relating to HDBTs:

- Potential U.S. adversaries worldwide are using underground facilities to conceal and protect their leadership, military and industrial personnel, weapons, equipment, and other assets and activities. These facilities include hardened surface bunkers and tunnel facilities deep underground.
- Many underground command, control, and communications (C^3) complexes and missile tunnels are between 100 and 400 meters deep, with the majority less than 250 meters deep. A few are as deep as 500 meters or even 700 meters, in competent granite or limestone rock.
- As identified by the Defense Intelligence Agency, there are about 10,000 HDBTs in the territory of potential U.S. adversaries worldwide.

- Of the 10,000 HDBTs identified, about 20 percent are estimated to have a major strategic function.
- Over half of these strategic HDBTs are located near or in urban areas.
- The number of known strategic HDBTs is increasing at a rate of about 10 percent per year. This increase is attributable mostly to discovery by the U.S. intelligence community and to a lesser extent to construction in countries seeking protection from U.S. military capabilities.
- With the current U.S. nuclear arsenal, a number of the more important strategic HDBTs cannot be held at risk of physical destruction of the functional area.
- A few hundred of the strategic HDBTs could be candidates for targeting with the robust nuclear earth penetrator (RNEP) weapon currently under study.

EXAMPLES OF STRATEGIC HARD AND DEEPLY BURIED TARGETS

Examples of strategic HDBTs are shown in Figure 2.1 and detailed in this section. Representative actual overburdens, not their reinforced concrete equivalents, are described in the following examples.

Missile Tunnel

Hard Target Type: Deep underground tunnel
Function: Deployment area for short-range ballistic missiles (SRBMs) that have a chemical weapon warhead; warhead mating performed in maintenance area
Site Location: Remote valley; nearest civilian population center is 30 kilometers away

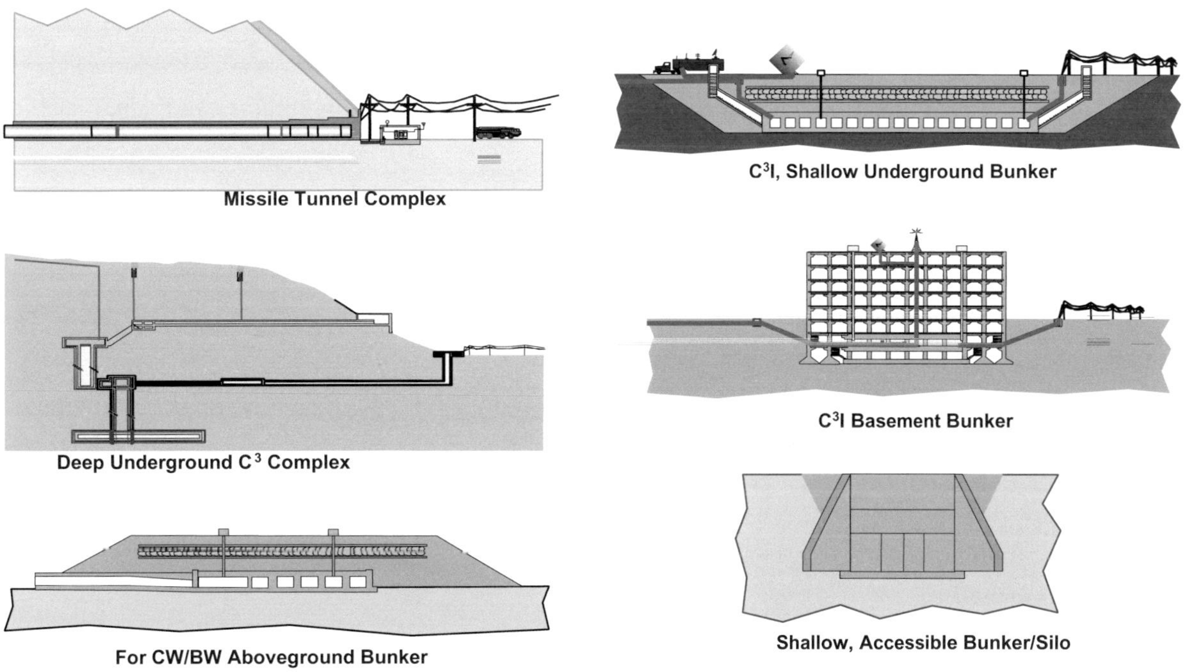

FIGURE 2.1 Examples of strategic hard and deeply buried targets. (See text for details. Acronyms are defined in Appendix D.)

Number of Stories or Levels: One

Number of Entrances: Two

Overburden: 120 meters at working area

Geology: Varying layer depths of residual soil, weathered rock, and limestone

Berms: Tunnel adits constructed to form a berm

Tunnels: Tunnel lining is 1 meter of reinforced concrete from entrance to blast doors, then 0.5 meter reinforced concrete throughout internal cavities

Blast Doors: 1 meter of steel-lined reinforced concrete

Reinforced Concrete: Compression strength of all concrete is estimated at 4,000 pounds per square inch

Deep Underground C^3 Complex

Hard Target Type: Bunkered facility deep underground

Function: Reserve post for providing command, control, and communications support to strategic nuclear forces as well as wartime protection for senior military authorities

Site Location: Mountainous region in country's interior, more than 200 kilometers away from a city of 50,000 and 6 kilometers from a village of 150

Number of Stories or Levels: Multilevel facility connected by elevators, shafts, and tunnels

Number of Entrances: Two horizontals entrances and one vertical shaft entrance

Overburden: 400 to 700 meters overburden at working area

Geology: Single monolithic upthrust of quartzite sandstone

Berms: None

Tunnels: Tunnel lining reinforced concrete

Blast Doors: Sliding double blast doors exist in excess of 0.7 meter thick at each entrance; main entrance is 7 meters wide, and auxiliary entrance is 9 meters wide

Reinforced Concrete: Reinforced concrete facing walls installed at both entrances, and a slab of reinforced concrete covering the vertical shaft

CW/BW Aboveground Bunker

Hard Target Type: Hardened aboveground bunker

Function: Chemical weapons (CW) munitions filling (artillery and bombs) and CW munitions storage

Site Location: Remote desert area, within a large chemical weapons production and R&D complex; no civilian facilities within a 10 kilometer range; nearest populated area is 40 kilometers northwest of the complex

Number of Stories or Levels: One

Number of Entrances: One

Overburden: 8 meters thick, including two 1 meter burster slabs separated by a 3 meter layer of 8 inch rocks, covered by soil

Ceiling: 1 meter of reinforced concrete
Walls: 1 meter of reinforced concrete
Floor: 2 meters of reinforced concrete
Roof: 0.5 meter of steel-lined concrete
Blast Doors: 0.5 meter of steel-lined concrete

C^{3}I Shallow Underground Bunker

Hard Target Type: Shallow underground bunker
Function: National-level military command-and-control facility
Site Location: Within a large military complex on the outskirts of a large city (population 250,000)
Number of Stories or Levels: One
Number of Entrances: Four
Overburden: Five layers above roof:
3 meters of compacted soil
1 meter of reinforced concrete burster slab
3 meters of crushed/compacted rock
1 meter of reinforced concrete burster slab
3 meters of compacted soil
Thickness of:
Walls: 1 meter of reinforced concrete
Floor: 1 meter of reinforced concrete
Roof: 2 meters of reinforced concrete
Blast Doors: 0.5 meter of steel-lined concrete

C^{3}I Basement Bunker

Hard Target Type: Basement bunker
Function: National-level command-and-control facility
Site Location: Within a military headquarters complex under a six-story Army Headquarters building in a heavily populated (750,000) urban area; civilian structures (hospitals, schools, embassies) all located within a 1 kilometer radius of the facility
Number of Stories or Levels: Two-level bunker underneath six-story building
Number of Entrances: Two
Thickness of:
Walls: Exterior, 2 meters; interior, 0.5 to 1 meter of reinforced concrete
Floor: 1 meter of reinforced concrete
Roof: 4.1 meters of reinforced concrete equivalent above upper bunker level
Blast Doors: 4 blast doors, sliding, steel-lined concrete

Shallow Accessible Bunker/Silo

Hard Target Type: Shallow buried "cut and cover" bunker
Function: Biological weapons (BW) agent storage and production
Site Location: Collocated within aboveground civilian bioproducts R&D and production complex; hilly terrain, vegetation; large civilian population within a 0.5 kilometer radius of the facility
Number of Stories or Levels: One
Number of Entrances: One
Overburden: 7 meters of compacted soil
Walls: 1 meter of reinforced concrete
Floor: 1 meter of reinforced concrete
Roof: 2 meters of reinforced concrete
Blast Doors: 0.5 meter overall; steel-lined concrete

3

Earth-Penetrator Weapons

An earth-penetrator weapon (EPW) is designed to detonate below the ground's surface after surviving the extremely high shock and structural loading that result during impact and penetration. As discussed in more detail in Chapter 4, detonating the weapon beneath the surface greatly increases ground-shock effects, making the weapon more effective in destroying hard and deeply buried targets (HDBTs).

EARTH-PENETRATOR TECHNOLOGY BACKGROUND

Earth-penetration technology in the United States dates to the early 1950s. The Mark 8, a nuclear bomb with the capability to penetrate soil and rock as well as concrete targets, entered the stockpile in January 1952. The Mark 11 bomb, a safety upgrade that replaced the Mark 8 in May 1957,[1] was removed from the stockpile in 1958.

Sandia National Laboratories (SNL) initiated an earth-penetration (EP) technology program in 1960.[2] The Department of Energy (DOE)/Department of Defense (DOD) programs, now the DOE's National Nuclear Security Administration (NNSA) national laboratories, and DOD laboratories have maintained continuous EP technology development programs and testing, at various levels of activity, since that time. Of the more than 3,000 EP tests conducted, there are currently 1,084 representative tests recorded in the SNL Earth Penetration Database. Complete characteristics of the penetrators—physical characteristics, impact velocity, impact angle, impact angle of attack, penetrator path length, penetrator rest angle, test location, target site, target material, date of test, and associated programs—are documented in this database. Geologic materials penetrated include sand, silt, and clay soils, frozen soil, ice, and rock. Penetration tests into concrete targets have also been conducted. Target sites were located in Alabama, Alaska, California, Florida, Kansas, Nebraska, Nevada, New Mexico, Texas, and Utah.

Important Parameters of Earth-Penetrator Weapons

Penetration tests have been conducted at various impact angles, angles of attack, and velocities into undisturbed geologic targets to provide insight into how the physical properties of a penetrator affect its performance. Data in the SNL database were obtained from weapon-sized penetrators instrumented to measure axial and lateral penetrator acceleration, strain in the EP case, and the structural response of internal components. Based on these data, Box 3.1 lists and defines physical properties of an EPW and impact conditions that are important in the design of an EPW. Figure 3.1 shows a typical EPW case configuration.

The ogive nose shown in Figure 3.1 is a ballistic shape that is formed by rotating an arc of radius R_o, tangent to the cylindrical body around the centerline of the body. A 6 caliber radius head (CRH) ogive nose is generally used on an EPW that is designed for the penetration of hard material at velocities of less than 900 meters per second. For higher velocities, a 3 CRH (blunter) is recommended in order to maintain penetrator stability. For impact velocities less than 300 meters per second into hard or frozen soil, a 9.25 CRH nose or a length-to-diameter ratio of 2 for cone-shaped-nose penetrators (sharper noses) can be used, since nose tip heating is not a problem at lower velocities. The flare on the rear of the penetrator in Figure 3.1 is important for penetrator stability if the length-to-diameter ratio of the EPW is less than 6.

BOX 3.1
Important Earth-Penetrator Weapon Parameters

Physical Characteristics
- *N:* nose shape
- *d:* body diameter, m
- *L:* total penetrator length, m
- *m:* total penetrator weight, kg
- *A:* cross-sectional area, m^2
- *m/A:* cross-sectional density, kg/m^2

Impact Conditions
- *V:* impact velocity, m/s
- θ: impact angle between velocity vector (trajectory angle) and target surface
- α: angle of attack, angle between velocity vector (trajectory angle) and earth-penetration axis

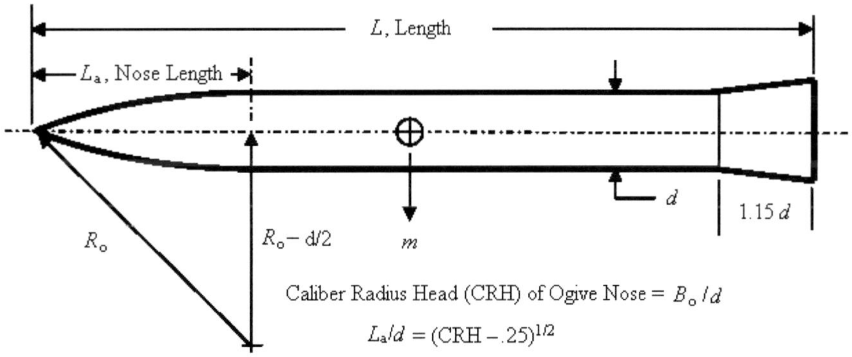

FIGURE 3.1 Typical earth-penetrator weapon case.

Earth-Penetrator Weapons Design Methodology

The national laboratories conducted two nuclear EPW engineering development programs between 1975 and 1990, using the SNL EP technology program design experiences and test database. The first program involved the W86 EPW payload for the Pershing II (P II) missile. Weapons-system-level EP tests that included warhead electrical system components and inert nuclear system components were conducted in soil, low-strength rock, and medium-strength rock at the Tonopah Test Range near Tonopah, Nevada. Missiles with system-level test units were also impacted into the 50 mile target area at White Sands Missile Range. The Los Alamos National Laboratory also conducted underground nuclear detonation tests of candidate nuclear systems. Engineering development of the W86 EPW was completed, and the design was put on the shelf. No weapons were produced.

In 1986 a feasibility study was completed on a strategic earth-penetrator weapon (SEPW) to hold deeply buried targets at risk. Shortly after the study was completed, an advanced development SEPW program was initiated. Los Alamos National Laboratory and Lawrence Livermore National Laboratory were tasked to design nuclear systems and hardware for penetration tests and to support underground tests. SNL was tasked to provide systems engineering support to each nuclear laboratory's effort. SNL was also tasked to design and develop the EP cases, hardened electrical components, and subsystems; to evaluate penetrator performance; and to investigate potential countermeasures. Approximately 60 penetration tests were conducted by the two laboratory teams. SEPW inert test units were impacted into a series of in situ (undisturbed) soil and rock targets at varying impact angles and velocities. A small number of underground nuclear tests were also conducted at the Nevada Test Site.

The programs described above are examples of programs in which nuclear weapons have been designed and developed to counter the uncertainties of an earth-penetration event and to enhance the survivability of an EPW. The important parameters of these weapon designs and the maximum impact velocity recorded in the SNL database are used in the calculations described below to predict the maximum depth of penetration that can be expected in soil, low-strength rock, and medium-strength rock media.

Target Geology

The greatest uncertainty in predicting EPW depth of penetration and structural survival of the weapon until detonation is due to the inherently heterogeneous nature of earth materials. Rock formations typically are composed of layers of materials of different strength such as the formations shown in Figure 3.2. They can also include joints and fractures as well as layers of different strength and sloping layers, as shown in Figure 3.3.

The type of massive, relatively homogeneous formation with few cracks and fissures that is shown in Figure 3.4 is rare. Even areas expected to have soil to extended depths may include areas of unexpectedly hard material. These uncertainties can be countered to some degree by designing an EPW to be as rugged as possible, consistent with mission and system requirements. Rugged EPWs have the highest cross-sectional density possible so as to enhance penetration depth, and a length-to-diameter ratio of 8 to 10 for stable trajectory; thus, lateral loading is minimized during penetration. A one-piece EP case fabricated from the best available high-strength, high-fracture-toughness steel is recommended in order to withstand the high lateral loading that occurs when the EPW encounters heterogeneous formations. Internal components and subassemblies must be designed and packaged to survive high-frequency structural loading. The maximum impact velocities and hardest expected target materials determine the selection of nose shape.

FIGURE 3.2 Limestone formations near Nashville, Tennessee, exhibiting layers of materials of different strengths. Photo courtesy of William J. Patterson.

FIGURE 3.3 Sandstone syncline in New Mexico showing sloping layers. Photo courtesy of William J. Patterson.

FIGURE 3.4 Massive sandstone in New Mexico showing relatively homogeneous rock. Photo courtesy of William J. Patterson.

EMPIRICAL EQUATIONS FOR PREDICTING PENETRATION CAPABILITY

Many equations predictive of earth-penetrating capabilities of EPWs have been developed over the past 40 years. In early 2000, two of the more widely used empirical equations were evaluated, and calculated depths were compared with large-scale tests of penetrators impacting into two different types of in situ rock formations at the Tonopah Test Range. Both of the equations predicted depths that agreed well with the Antelope Tuff and the Sidewinder Welded Tuff rock penetration data documented in the SNL Earth Penetration Database.[3] One equation, developed by C.W. Young, was published in 1967.[4] The most recent update was in 1997.[5] The other equation was developed by M.J. Forrestal and published in 1994.[6]

Presented here (Box 3.2), Young's equation illustrates the use of EP parameters to calculate maximum depths of penetration achievable in soil, low-strength rock, and medium-strength rock, as shown in Table 3.1.

The following sample calculation with Young's equation uses the SEPW parameters and an impact velocity of 1,220 meters per second in low-strength rock.

BOX 3.2
Young's Empirical Equation

$$D = \alpha K_s SN(m/A)^{0.7}(V_s - 30.5)$$

Where D = depth of penetration in meters, α = 0.0000175, K_s = scaling factor, S = empirical target constant, N = penetrator nose coefficient, m = penetrator mass in kilograms, A = penetrator cross-sectional area in square meters, and V_s = impact velocity in meters per second.

For soil the scaling factor K_s is defined as

$$K_s = 1.0 \text{ if } m \geq 27 \text{ kilograms}$$

and

$$K_s = 0.27(m)^{0.4} \text{ if } m < 27 \text{ kilograms.}$$

For rock and concrete the scaling factor K_s is defined as

$$K_s = 1.0 \text{ if } m \geq 182 \text{ kilograms}$$

and

$$K_s = 0.46(m)^{0.15} \text{ if } m < 182 \text{ kilograms.}$$

The nose coefficient N for a tangent ogive is defined as

$$N = 0.18(L_n/d) + 0.56,$$

while the nose coefficient for a conical nose is defined by

$$N = 0.26(L_n/d) + 0.56.$$

In the nose coefficient equations, L_n is the length of the penetrator nose in meters, and d is the diameter of the penetrator body in meters.

The constant S is an empirical value, which depends on the target material. Its value is determined by measuring the depth of penetration for a given penetrator tested in a given geologic material and applying the equation to compute the value of S. A reasonable value of the empirical constant for a given geologic material is obtained by averaging several of the values for S obtained from several penetration tests. Once S has been determined experimentally for a given target material, the other variables in the equation may be altered to estimate depth of penetration into the same target material.

TABLE 3.1 Empirically Estimated Maximum Credible Depths of Penetration of Earth-Penetrator Weapons in Three Types of Rock and Soil Media

Weapon	Rock/Soil Classification	*S* No.	CRH	*N*	*m*, kg	*d*, m	*m/A*, kg/m^2	*V*, m/s	Penetration Depth: *D*, m	a_{ave}	a_{peak}
Strategic EPW (SEPW)	Medium-strength rock	0.76	3	0.85	411	0.27	7,010	700	3.7	6,702	10,000
Low-yield EPW	Medium-strength rock	0.76	6	1.00	184	0.17	8,451	1,000	7.2	7,045	10,000
Optimized EPW	Medium-strength rock	0.76	6	1.00	2,668	0.37	25,044	500	7.5	1,700	2,500
SEPW	Low-strength rock	1.30	3	0.85	411	0.27	7,010	1,200	11.1	6,592	10,000
Low-yield EPW	Low-strength rock	1.30	6	1.00	184	0.17	8,451	1,500	18.7	6,114	9,000
Optimized EPW	Low-strength rock	1.30	6	1.00	2,668	0.37	25,044	500	12.8	994	1,500
SEPW	Silty clay	8.00	3	0.85	411	0.27	7,010	1,500	86	1,332	2,000
Low-yield EPW	Silty clay	8.00	6	1.00	184	0.17	8,451	1,500	115	993	1,500
Optimized EPW	Silty clay	8.00	6	1.00	2,668	0.37	25,044	500	79	161	250

Notes: *S* no. = empirical target constant; CRH = caliber radius head; *N* = penetrator nose coefficient; *m* = penetrator mass; *d* = diameter of penetrator body; *m/A* = cross-sectional density; *V* = impact velocity; *D* = depth of penetration; a_{ave} = average acceleration; a_{peak} = maximum acceleration.

Given

$$D = \alpha K_s SN(m/A)^{0.7}(V_s - 30.5)$$

where $\alpha = 0.0000175$, $S = 1.3$, nose CRH = 3, $m = 411$ kg, $d = .274$ m, and $V = 1{,}220$ m/s, then $L_n/d = (\text{CRH} - .25)^{1/2} = (3 - .25)^{1/2} = 1.66$, $N = (.18 \times 1.66) + .56 = .85$, $A = \pi(.274)^2/4 = .059\ \text{m}^2$, and $(m/A)^{0.7} = (411/.059)^{0.7} = 492$.

Hence

$$D = 0.0000175 \times 1.3 \times .85 \times 492 \times (1{,}210 - 30.5)$$
$$= 11.3 \text{ m, estimated minimum penetration depth.}$$

Based on data for deceleration versus time from a large number of tests, the best estimate of peak rigid body deceleration is 1.5 times the average deceleration:

$$a_{ave} = V^2/2gD = 6{,}646\ g \text{ and } a_{peak} = 1.5 \times 6{,}646 = 9{,}968\ g.$$

MAXIMUM CREDIBLE DEPTHS OF PENETRATION

Table 3.1 presents empirically estimated maximum depths of penetration in three typical geologic materials. These depths were calculated using Young's empirical equation. In order to add credibility to these estimates, the following data were used in the calculations: the physical properties of the W86 P II EPW and SEPW designs, penetrability numbers for the geologic media supported by test data, and impact velocities no greater than the highest velocities documented in the SNL Earth Penetration Database. The results of an EPW bomb optimized with the highest *m/A* feasible, within a 2,700 kilogram weight limit, are also shown to illustrate the ability of a robust EPW to minimize axial deceleration.

The 10,000 *g* peak deceleration capability of the SEPW and W86 P II EPW was the limiting factor in medium-strength rock. The impact velocity used in the calculations in low- and medium-strength rock was limited by the 10,000 *g* peak deceleration capability of the SEPW and W86 P II EPW. Also, impact velocity was limited to 1,525 meters per second, since no test data exist above 1,525 meters per second. The heavy weight of the EPW bomb limits the delivery system to aircraft only; therefore, the impact velocity for the EPW bomb was based on a reasonable maximum velocity obtainable from a high-altitude airdrop.

The heavy, high-cross-sectional-density bomb impacting at 500 meters per second achieved the maximum calculated depth in medium-strength rock, and the peak axial deceleration did not exceed 2,500 *g*. The low-yield EPW impacting at 1,500 meters per second achieved the maximum depth in low-strength rock and silty clay soil. Peak axial decelerations were no greater than 9,000 *g* and 1,500 *g*, respectively.

It must be kept in mind that these calculations assume that the penetrated medium is homogeneous; thus, these are the minimum expected depths. The maximum depth in soil could vary by ±20 percent based on the accuracy of Young's equation.[7] This would give a maximum depth of penetration for the low-yield EPW of approximately 140 meters in soil. The depths in rock could show even greater variability due to the nonhomogeneity of rock formations. Depths up to 50 percent greater than pretest estimates have been observed. The maximum depth in low-strength rock could be approximately 30 meters, and approximately 12 meters in medium-strength rock. Designers and test engineers are usually most interested in estimates of minimum depth since minimum depth results in the highest expected axial EPW deceleration.

W61 SYSTEM

The W61 EPW was intended to be an interim weapon to provide an EPW capability until the SEPW was fielded. The B61-7 was selected for conversion into an EPW because its internal components were required to survive relatively high axial and lateral acceleration loads. The B61-7 had to survive loading from an impact velocity of approximately 30.5 meters per second onto hard surfaces—impact conditions resulting from low-level, high-speed aircraft release and parachute-retarded lay-down. Detonation time was set for a safe aircraft separation time.

W61 predevelopment engineering began in 1987. Engineering development began in 1990, and the program was terminated in 1992. The W61 was designed to be a rapidly deployable system. The delivery system had the capability of delivering the W61 to the target surface at optimum impact conditions. For the targets of interest, optimum conditions were impact angles within 10 degrees of target normal, an angle of attack (angle between velocity vector and EPW centerline) of less than 2 degrees, and impact velocities around 245 meters per second.

Following are the initial basic guidelines from the DOD for the conversion of the B61-7 bomb into a W61 EPW:[8]

- Minimize new component development,
- Maximize the use of existing hardware, and
- Minimize changes to the B61-7 electrical system.

A one-piece, cone-nosed penetrator case was designed to house the nuclear system and the warhead electrical system. The case was fabricated from high-strength steel with high fracture toughness. Since changes to components of the nuclear system and warhead electrical system were not allowed, the W61 EPW survivability limit was governed by the deceleration capability of the internal components. The system survivability level was determined by testing. Tests conducted during engineering development demonstrated that the W61 had the capability to survive penetration of 0.3 meter of concrete, hard soil, and low-strength rock at specified impact conditions. The W61 had an airburst and a contact-burst capability as well as subsurface-burst capability.

B61-11 EARTH-PENETRATING BOMB

The B61-11 was developed to replace the B53 gravity bomb, which had entered the stockpile in 1962. In 1988 an interim nuclear safety modification was made, resulting in the B53-1. However, even with the modification, the B53 did not completely meet standards for modern weapons safety, security, and reliability. Figure 3.5 and Table 3.2 show the final design and delineate the properties of the B61-11.

Following are the initial basic guidelines from the DOD for conversion of the B61-7 into an earth-penetrating bomb and the resulting actions:

- Carry out a rapid development program.
  —The program was authorized on September 15, 1995.
  —Major assembly release occurred on December 30, 1996.
- Minimize new component development.
  —EP case forgings from the cancelled W61 program were used.
  —No changes were made to the B61-7 nuclear system.
  —No changes were made to the B61-7 electrical system.

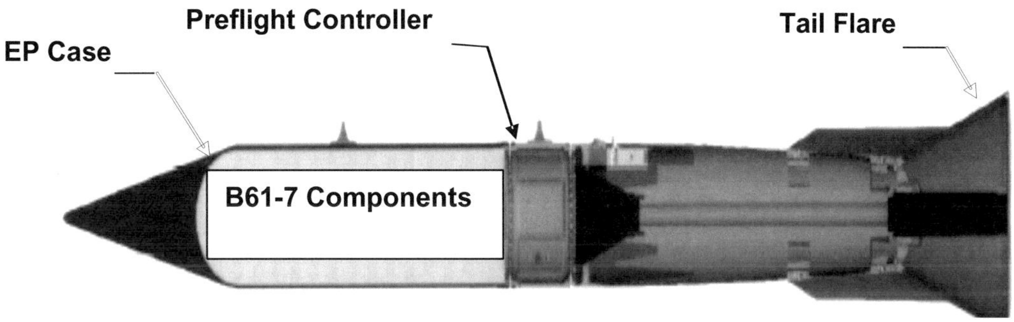

FIGURE 3.5 Final design of the nuclear, earth-penetrator (EP) B61-11. SOURCE: Sandia National Laboratories. 2003. *B61-11 Final Weapon Development Report (U),* SAND 2003-2344, Albuquerque, N.Mex. (Classified).

TABLE 3.2 Comparison of B61-7 and B61-11 Properties

Property	B61-7	B61-11
Mass	346 ± 7 kg	549 ± 7 kg
Nose, aluminum center case, and earth-penetrator case	243 kg	377 kg
Preflight controller	22.5 kg	22.5 kg
Tail	80.7 kg	150 kg
Diameter	0.33 m	0.34 m
Length	3.6 m	3.7 m

Because changes to the nuclear system and warhead electrical system were not allowed, the survivable deceleration level of the B61-11 was limited to the deceleration limits of the B61-7 internal components. Penetration tests with functional warhead electrical-system components and simulated nuclear assemblies were conducted at different impact velocities into hard soil and frozen soil to demonstrate B61-11 capability in the targets of interest.

The classified military requirements for the B61-11 include limits for soil penetration capabilities, yield, center of gravity, reliability, stockpile quantities, and ballistic characteristics.

The B61-11 was developed to ensure a capability to continue to hold selected deeply buried targets at risk.

CURRENT ROBUST NUCLEAR EARTH PENETRATOR PROGRAM

The Robust Nuclear Earth Penetrator program is an engineering feasibility study. It was initiated in May 2003 with the intention of its being a 2-year study, with two teams working on the study: a Los Alamos National Laboratory (LANL) and Sandia National Laboratories Albuquerque (SNLA) team, and a Lawrence Livermore National Laboratory (LLNL) and Sandia National Laboratories Livermore (SNLL) team. The purpose of the program is to determine if, using the major components of an existing weapon system, an earth-penetrator system can be designed that can hold at risk a significantly larger number of targets than the B61-11 can. The LANL/SNLA team's focus is the B61-7, and the LLNL/SNLL team is addressing the B83. The basic approach is to increase the cross-sectional density of the

EPW when possible and to increase the deceleration capability level of the internal components. No changes in the parent weapon yield are allowed. The study is allowed to address potential changes to internal components, as long as the changes do not require any nuclear certification tests. Owing to budget constraints the study is now limited to the robust nuclear earth penetrator (RNEP) being studied by the LLNL/SNLL team. At this time there is no decision on when or if the LANL/SNLA team will restart its study.

Following are general guidelines from the DOD for the RNEP under study by the LLNL/SNLL team:[9]

- The RNEP weapon is required to be able to do the following:
  —Survive penetration and not rebound from the target;
  —Reach a certain depth in a specified geology (the "threshold");
  —Preserve or improve original weapon functionality; and
  —Preserve or improve original weapon safety, security, and reliability.
- Regarding the compatibility of the RNEP with delivery aircraft:
  —The maximum weight, length, and diameter of the EPW case are to be determined by delivery aircraft requirements.
- Modifications to the Arming, Fusing, and Firing (AF&F) system are allowed.
  —The AF&F capability level is to be determined by structural testing.
- Modification of the nuclear system is allowed provided no nuclear certification testing is required.

The properties of the LLNL/SNLL RNEP are as follows:

- Mass, including tail kit—1,379 kg,
- Diameter—0.53 m, and
- Penetrator length (may be extended by tail kit)—2.54 m.

Following are guidelines from the DOD regarding the RNEP guidance system:

- RNEP is to be a guided and controlled weapon system. Its guidance system will do the following:
  —Allow for precise targeting,
  —Allow for optimization of angle of attack and incidence control, and
  —Minimize the stresses on the EPW system.
- RNEP system capability will be determined by experimentation, test, and analysis.
  —The RNEP survivability level is set by the structural limit of the nuclear system and the arming, fusing, and firing system.

NOTES

1. F.C. Alexander. 1967. *History of Gun Type Bombs and Warheads Mark 8, 10, 11 (U)*, SC-M-67-658, Sandia National Laboratories Library, Albuquerque, N.Mex., May (Classified).

2. W.N. Caudle and A.Y. Pope. 1962. *Project Trump: Progress Report No. I*, SCTM 56-62 (71), Sandia Program for Earth Penetrating Systems, Sandia Corporation, Albuquerque, N.Mex., April.

3. W.J. Patterson and R.S. Baty. 2003. "Comparison of Two Empirical Equations with Large Scale Penetrator Tests into In Situ Rock Targets," 11th International Symposium on Interaction of the Effects of Munitions with Structures, Mannheim, Federal Republic of Germany, May 5-9.

4. C.W. Young. 1967. *The Development of Empirical Equations for Predicting Depth of an Earth Penetrating Projectile*, SC-DR-67-60, Sandia National Laboratories, Albuquerque, N.Mex., January.

5. C.W. Young. 1997. *Penetration Equations*, Contractor Report, SAND 97-2426, Sandia National Laboratories, Albuquerque, N.Mex., October.

6. M.J. Forrestal, B.S. Altman, J.D. Cargile, and S.J. Hanchak. 1994. "An Empirical Equation for Penetration Depth of Ogive-Nose Projectiles into Concrete Targets," *International Journal of Impact Engineering,* Vol. 15, No. 4, pp. 395-405.

7. W.J. Patterson and R.S. Baty. 2003. "Comparison of Two Empirical Equations with Large Scale Penetrator Tests into In Situ Rock Targets," 11th International Symposium on Interaction of the Effects of Munitions with Structures, Mannheim, Federal Republic of Germany, May 5-9.

8. Sandia National Laboratories. 1992. *W61 Weapon Development Report (U),* SAND 91-2243, RS 3151/91/00024, Albuquerque, N.Mex., March 1 (Classified).

9. Sandia National Laboratories. 2003. *B61-11 Final Weapon Development Report (U),* SAND 2003-2344, Albuquerque, N.Mex. (Classified).

4

Effectiveness of Nuclear Weapons Against Hard and Deeply Buried Targets

TARGET DESTRUCTION

The types of hard and deeply buried targets (HDBTs) of interest are described in Chapter 2. These range from hardened, surface bunker complexes to tunnel complexes deep underground, as shown in Figure 2.1.

Nuclear weapons are the only weapons that can destroy targets deep underground or in tunnels. Weapons of mass destruction (WMD)-related facilities near the surface may be destroyed with either nuclear or conventional weapons, but nuclear weapons are of interest because they produce more effective agent-kill mechanisms. The types of conventional weapons likely to be employed against tactical targets are discussed in Chapter 7, "Conventional Weapons." Note that the discussion in this chapter assumes the physical destruction of the target. It may also be possible to destroy or degrade the functionality of the facility, or the functionality of the network of which it is a part, without physically destroying the specific target node.

The following are elements of target destruction:

- Finding, identifying, and characterizing the target;
- Weapon-system survival and arrival at the target;
- Weapon penetration and detonation;
- Energy coupling of weapons effects to the ground;
- Shock propagation through the ground to the target facility; and
- Response and vulnerability of the target facility.

Finding, Identifying, and Characterizing the Target

This report examines the relative effectiveness of and collateral damage associated with nuclear earth-penetrator and other weapons. It is not within the scope of the committee's tasks to analyze the quality of the intelligence that exists or will exist. Nor does the committee discuss how good the intelligence must be to enable the effective operational use of a weapon.

The discussions and calculations presented in this report assume knowledge of the target's location, purpose, size, function, internal layout, and other relevant features at the time it may be attacked. This assumption of perfect, timely intelligence is unlikely to hold in reality for the vast majority of targets of interest.

Important to the issue of finding, identifying, and characterizing a target is that, in addition to concealment, deceptive techniques are used extensively by adversaries to complicate matters. The calculations presented should all be viewed with these intelligence uncertainties in mind.

Weapon-System Survival and Arrival at the Target

In addition to the passive defense of hardening a target, high-value sites are generally defended. Thus, the method of delivery of a weapon is important. For example, defenses against a weapon that is air-delivered are quite widespread, whereas ballistic missile defenses are virtually nonexistent. This report does not examine the effectiveness of delivery modes or defenses against them.

Weapon Penetration and Detonation

There is the need to consider the probability that an earth-penetrator weapon (EPW) will survive ground penetration, penetrate to the desired depth, and then successfully detonate. In comparison with surface (contact) burst weapons, the EPW experiences more rigorous impact conditions. These factors are discussed in Chapter 3.

In the following discussion of weapons effects at depth, the committee presumes that the nuclear earth-penetrator weapon successfully reaches the target, penetrates any aboveground covering structure and possible defenses, and enters the surface to a depth sufficient to couple the majority of its energy to the ground, all with a probability of 1.0. The committee did not study the probability of any of these events.

Energy Coupling of Nuclear Weapons Effects to the Ground

The energy coupled by a nuclear weapon to the ground is expressed as the fraction of the total weapon yield converted to kinetic energy of downward-moving solid or nonvaporized ground material. The amount of energy coupled to the ground is strongly dependent on the weapon's actual height of burst (HOB) or depth of burst (DOB), as well as on nuclear design details (i.e., yield-to-mass ratio, fission fractions, and relative coupling efficiencies of the source components). Geologic properties also play a role.

Effects Manual-1: Capabilities of Nuclear Weapons[1] of the Defense Threat Reduction Agency (DTRA) (formerly the Defense Nuclear Agency) defines an equivalent yield factor for both total coupled energy and ground-shock-coupled energy as a function of HOB/DOB (see Appendix C in this report for details). Figure 4.1 shows the equivalent yield factors normalized to a contact burst using the DTRA-recommended, scaled HOB of 0.05 m/kt$^{1/3}$. Note that the coupled energy is not defined for a scaled HOB greater than –0.05 m/kt$^{1/3}$ or a scaled DOB greater than 0.05 m/kt$^{1/3}$ due to uncertainties in calculations for this near-surface region. The ground-shock-coupled energy[2] includes surface air-blast-induced ground shock.

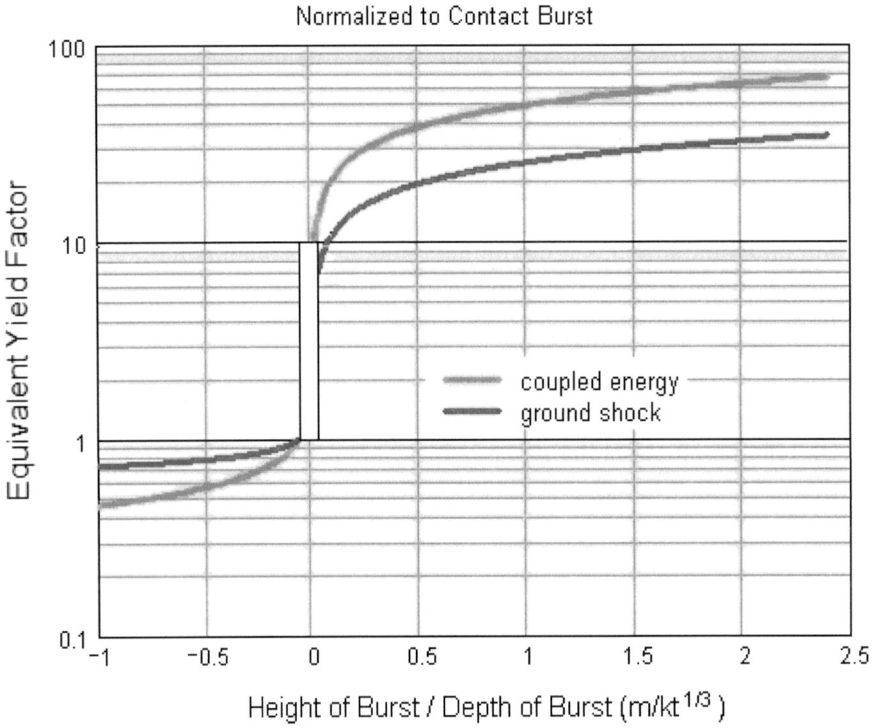

FIGURE 4.1 Equivalent yield factors for total coupled energy and ground-shock-coupled energy normalized to a contact burst. Positive numbers signify below ground. (See Appendix C.)

Equivalent yield factors for coupled energy asymptotically approach 100 (i.e., as burial depth increases); indeed an earth-penetrator weapon is effectively fully coupled at a scaled DOB of about 2.3 m/kt$^{1/3}$. Relative to a contact burst, the ground-shock-coupling factor approaches 50 with increased DOB owing to the surface air-blast contribution.[3]

For a generic 300 kiloton EPW at 3 meters depth of burst (scaled DOB = $3/(300)^{1/3} = 0.45$), the ground-shock-coupling factor is about 20, which is equivalent to a contact burst of about 6.0 megatons.[4] This example illustrates the "efficiency" of an earth-penetrator nuclear weapon in generating comparable levels of damaging ground shock at target depth with significantly lower yield relative to a surface-burst or airburst weapon. As mentioned elsewhere, the coupling factor can be anywhere between 15 and 25, with the greatest uncertainty due to the effect of the radiation from a surface burst, which is sensitive to local conditions.

Uncertainties in Weapon Effectiveness, Energy Coupling, and Ground-Shock Range to Effect

The following factors influence the uncertainties associated with estimating energy coupling for ground shock.[5]

- *Weapon design.* Energy deposition into the ground involves two components, which couple differently: debris kinetic energy and x-rays. The physical dimensions and the location of the primary component relative to the secondary also affect energy deposition.

- *Actual height of burst (HOB) or depth of burst (DOB).* Coupling increases by more than an order of magnitude in the region between contact and 1 m/kt$^{1/3}$ DOB. Defining "contact" is an issue; for example, a bomb's "center of energy" does not fully describe the orientation of a staged device. The size of a hot, nuclear source is also important, since the absorption of radiation does not scale with yield. DTRA has adopted 0.05 m/kt$^{1/3}$ as a standard value for the HOB of a contact (nose-down) burst to avoid elaborate numerical simulations of the energy-coupling process for each weapon type and geology.
- *Site geology and damage criterion.* The type of geologic media in which the source energy is deposited (dry/wet soil/rock) affects coupled energy. Reflections and/or rarefactions occur as the ground shock propagates and impinges on stiffer or softer geologic layers. Layering can significantly influence the range to effect for a particular criterion of damage (i.e., peak overstress, peak particle velocity, or peak free-field strain). The influence of the lethality criterion is discussed below (see the subsection entitled "Response and Vulnerability of the Target Facility").

To be fully contained (i.e., with no venting), a 300 kiloton weapon would need to be buried about 800 meters[6,7] and the emplacement hole would need to be carefully stemmed.[8] Because the practical penetration depth for an EPW is but a small fraction of the depth for full containment and the penetration hole would not be stemmed, there will be surface venting, prompt and residual nuclear radiation, and fallout effects from an EPW. For maximum energy coupling, analysts are most interested in the maximum depths. As can be seen in Table 3.1, which shows maximum empirically estimated EPW depths, none of the depths is great enough to contain an EPW nuclear burst, even if the penetration hole is stemmed. These effects are discussed later in this chapter.

There is a reasonably extensive experimental database, *Effects Manual-1 (EM-1),*[9] covering the various physics regimes governing the energy-coupling process. Analytical tools are fairly well advanced also. Uncertainties associated with estimates of energy coupling are far greater for near-surface airbursts than for buried bursts and depend on how well the actual burst location and details of weapon energy output are known. For present purposes with the bomb design fully prescribed and a DOB of ≥3 meters assumed, the effective ground-shock-producing yield can be reliably estimated to within about 20 percent. The greatest uncertainties in yield equivalency are in the immediate vicinity of the ground surface, where coupling is ill-defined at the actual air-ground interface (shown in the region ±0.05 m/kt$^{1/3}$ in Figure 4.1). Uncertainties exceed a factor of 2 for bursts of small heights where, of course, the ground-coupled energy is considerably less.

Shock Propagation Through the Ground to the Target Facility

Shock propagation through the ground to the target facility depends on weapon yield and coupled energy, stratigraphy of the target site, and properties of the intervening geologic materials, including joints and fault patterns. The ground-shock environment transmitted to the target facility is described in terms of material stresses, strains, particle velocities, and displacements (time-dependent details and peak values), as well as discontinuous (block) motions that can occur along joint and fault surfaces.

DTRA and the Department of Energy (DOE) weapon laboratories have invested considerable resources over the years to develop computational methods for predicting the ground-shock environments at depth from both high-explosive and nuclear bursts. The problem is complicated owing to various shock attenuation mechanisms, such as inelastic effects, hysteresis, and fracture and dilatation, and to geometric effects due to divergence, layering, interfaces, faults, and joints. Substantial progress has been made in the modeling of these processes, and there have been a number of successful predictions of large-scale, high-explosive events as well as underground (cavity) nuclear events. However, these correlations were obtained for tests performed in media that were easily accessible for extensive pretest measurement of properties. This will rarely, if ever, be the case in targeting applications for which limited site geologic descriptions are only imprecisely known at best. Moreover, significant changes in geologic properties can occur locally as well. Thus, estimates of peak velocity or stress at depth should not be expected to be accurate to better than a factor of 2.

Response and Vulnerability of the Target Facility

Assessment of the response and vulnerability of the target facility introduces still additional uncertainty beyond that already discussed. Generally speaking, the strength of the target depends on the type, physical properties, and quality (jointing) of the surrounding geology; the method of construction and shock orientation; and the design of any internal liners, including those for possible shock isolation of internal components. The tunnel appears to be stronger under "end-on" loading (when the shock strikes at a glancing angle to the tunnel's longitudinal axis) than when the tunnel is struck side-on.

All of the schemes for predicting tunnel damage are underpinned by a limited applicable experimental database of eight underground nuclear tests (UGTs) in which tunnels of various construction types were exposed to damaging ground-shock levels in a few types of rock geologies. The liner construction illustrated in the lower right of Figure 4.2 refers to various levels of protection, the least being reinforcement of the bare tunnel walls with rock bolts and wire mesh (an extension of what typically is done for safety considerations), and the greatest being either a ductile internal liner (composite lining) or an internal liner back-packed with a crushable material. Typical hardness ranges are indicated below each of the tunnel cross-sectional diagrams. Only two of these tests, Hard Hat (1962) and Pile Driver (1966), were dedicated experiments on structures in competent granite geology. The others were add-on experiments to UGTs conducted for different purposes in relatively soft tuff geology.

A summary of this experiment database is shown in Figure 4.2, which lists the UGT events and the number and types of structures tested. These data are limited to relatively small, lined and unlined tunnel sections, in either jointed hard rock (granite) or unjointed soft rock (tuff). All of the Hard Hat structures were 2 meters in interior diameter, while the Pile Driver structures ranged from 1 to 6 meters in interior diameter. The diameter of the tunnel excavation prior to construction of the liners ranged from 2 to 15 meters in both tests. In the major soft-rock tests (Mighty Epic, 1976, and Diablo Hawk, 1978) all lined tunnel and capsule structures were 1 meter in interior diameter. Typical excavated diameters were 2 meters. Also tested were 2-meter-interior-diameter X- and T-intersections connecting to 1-meter-interior-diameter tunnel stubs.

All of the experimental tunnels experienced severe damage at peak stress levels of about 1 to 1.5 kilobars or less, with the exception of some extremely hardened tunnels in granite that survived stress levels of 3 to 4 kilobars. More specifically, in the hard-rock tests, all structures in excavations of 5 meters or greater completely collapsed. A spectrum of damage occurred for structures in smaller-diameter excavations. It should be noted that the objectives of the hard-rock experiments were to develop survivable designs, and thus a number of the construction techniques investigated were of a more heroic design than would be likely for current targets. Thus, the conservatism inherent in the test designs and, to an extent, in the interpretation of results is not conservative from the perspective of vulnerability assessment. Nonetheless, the design and vulnerability assessment methods derived from this test experience underlie many of today's target-planning procedures (e.g., determination of the ground vulnerability number (GVN) or the characterization of the hardness of underground facilities by the Defense Intelligence Agency).[10]

The DTRA contractor community and the DOE weapons laboratories have advanced analytical

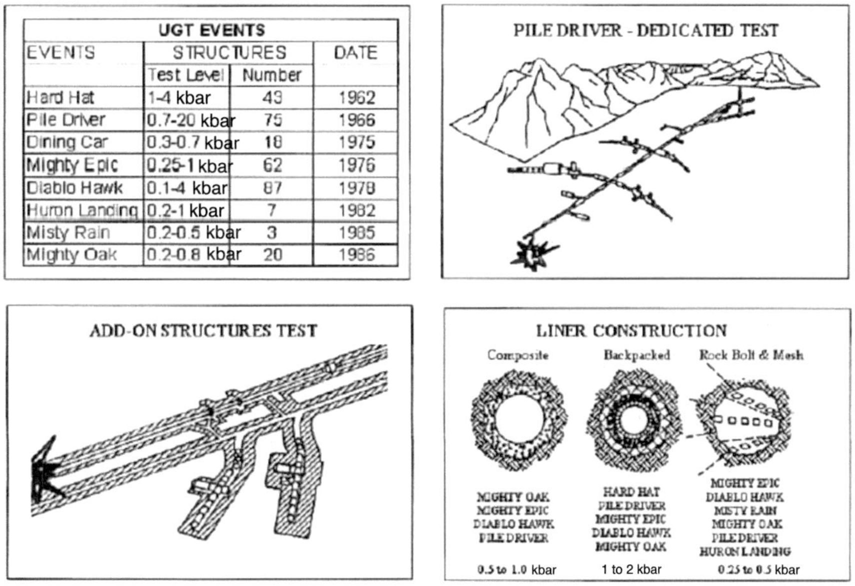

UGT EVENTS			
EVENTS	STRUCTURES		DATE
	Test Level	Number	
Hard Hat	1-4 kbar	43	1962
Pile Driver	0.7-20 kbar	75	1966
Dining Car	0.3-0.7 kbar	18	1975
Mighty Epic	0.25-1 kbar	62	1976
Diablo Hawk	0.1-4 kbar	87	1978
Huron Landing	0.2-1 kbar	7	1982
Misty Rain	0.2-0.5 kbar	3	1985
Mighty Oak	0.2-0.8 kbar	20	1986

FIGURE 4.2 Database for Underground Nuclear Test (UGT) Survivability/Vulnerability (U). SOURCE: Reprinted with permission from Defense Science Board, *Report on Underground Facilities,* June 1998.

capabilities for assessing the vulnerability of hardened tunnels. All are calibrated in one manner or another to the applicable experimental database, as noted earlier. The use of large-scale, finite-element structural dynamic codes is commonplace. Damage typically is associated with the degree of tunnel closure (i.e., maximum circumferential strains in the tunnel liner) calibrated to the Hard Hat/Pile Driver test data. Currently, DTRA is developing an improved GVN methodology that distinguishes between crushing of the tunnel wall (and liner), a so-called global damage mode, and more local rock spall, referred to as a local damage mode. This latter damage mode is controlled by the normal component of velocity at the rock surface and is calibrated against tests of relatively shallow tunnels loaded by high-explosive charges. It is mostly applicable to unlined or minimally lined tunnels.

As part of DTRA's program for improving GVN methodology, the global damage mode has been recalibrated to virtually the entire underground nuclear test database (i.e., *EM-1*) in terms of peak free-field strain in the surrounding rock, rather than peak stress, as was done in the older GVN methodology. This more physically attractive approach is made possible by modem computational capabilities. An ongoing DTRA experimental program (the Target Tunnel Defeat Advanced Concept Technology Demonstration) is intended to verify the improved vulnerability assessment method by means of a series of tunnel experiments at various scales in jointed limestone geology, the final test being of a prototype tunnel configuration loaded by a high-explosive simulation of a low-yield EPW. This test is scheduled for late 2005.

Ground-Shock Attenuation with Depth

The effectiveness of nuclear weapons against deeply buried targets can be estimated by calculating the intensity of the ground shock in the vicinity of the buried target in relation to target hardness.

Ground-shock calculations using DTRA's state-of-the-art, two-dimensional, physics-based ground-shock code, WinGS, were done for a 5.6 megaton contact burst and a 300 kiloton EPW at 3 meters' depth of burst in a representative homogenous granite target site; the ratio in yields was determined from the energy-coupling relationship, as discussed previously. The attenuation of peak free-field strain, stress, and velocity with depth below the ground surface resulting from these calculations is shown in Figures 4.3(a), (b), and (c), respectively. Peak stress contours for the two weapon types are compared in Figure 4.4. The general agreement between the results for the two weapon types (EPW and contact burst) supports the equivalence based on energy coupling.

Range to Effect

The hardness of tunnel-type target facilities of interest is expressed in terms of peak free-field strain (global damage mode) or peak free-field velocity (local damage mode). For an unlined or modestly lined tunnel (e.g., rock bolts and reinforced concrete liners) of 10 to 20 meters in diameter in the representative granite site, a 50 percent probability of severe damage (i.e., tunnel collapse) occurs at peak strains of 0.15 to 0.20 percent and velocities of 5 to 15 meters per second. As indicated in Figures 4.3(a) and (b), the range to effect for these hardness criteria is about 400 meters. At this range, target hardness expressed in terms of peak free-field stress is about 1 kilobar. Thus, either the 5.6 megaton contact burst or 300 kiloton EPW at 3 meters' depth of burst can drive damaging levels of ground shock to depths of around 350 to 400 meters (range to effect) in a competent granite site. These estimates do not take into account weapon delivery accuracy (circular error probable [CEP]). As discussed later in this chapter, the probability of severe damage at these depths is about 0.5 for a 10 meter CEP. For higher probabilities of damage, say 0.95, the range to effect is reduced to around 250 meters for a 10 meter CEP, or 150 meters for a 100 meter CEP (see Figure 4.6 in the following section).

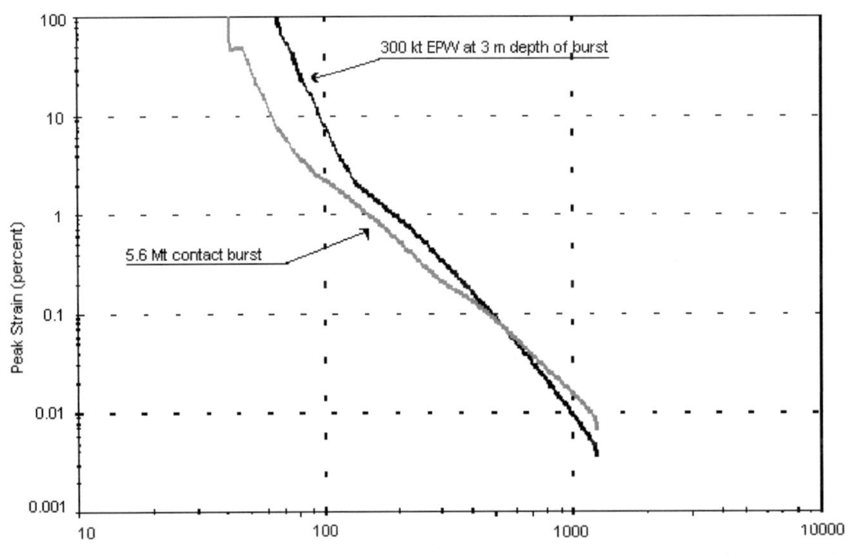

FIGURE 4.3(a) Peak strain versus target depth for 300 kt earth-penetrator weapon (EPW) and "damage equivalent" 5.6 Mt contact burst.

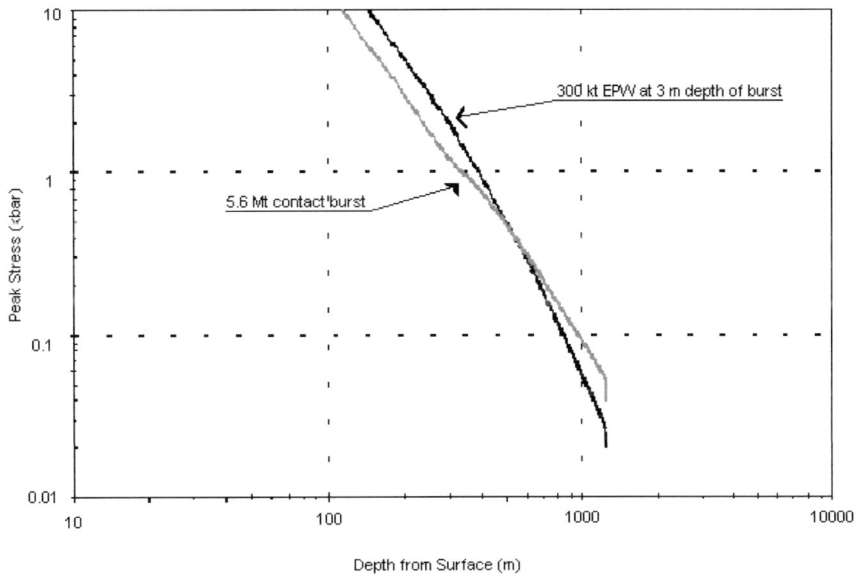

FIGURE 4.3(b) Peak stress versus target depth for 300 kt earth-penetrator weapon (EPW) and "damage equivalent" 5.6 Mt contact burst.

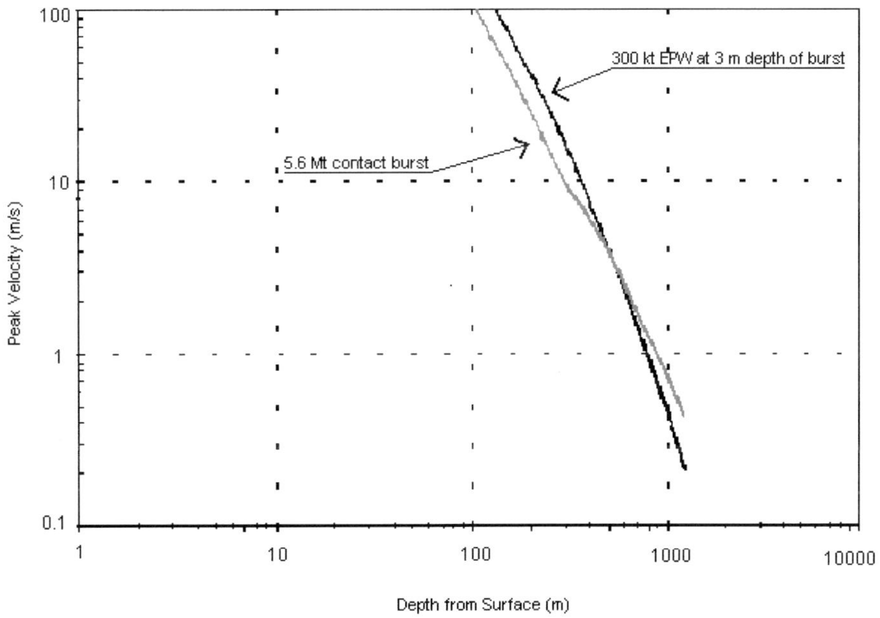

FIGURE 4.3(c) Peak velocity versus target depth for 300 kt earth-penetrator weapon (EPW) and "damage equivalent" 5.6 Mt contact burst.

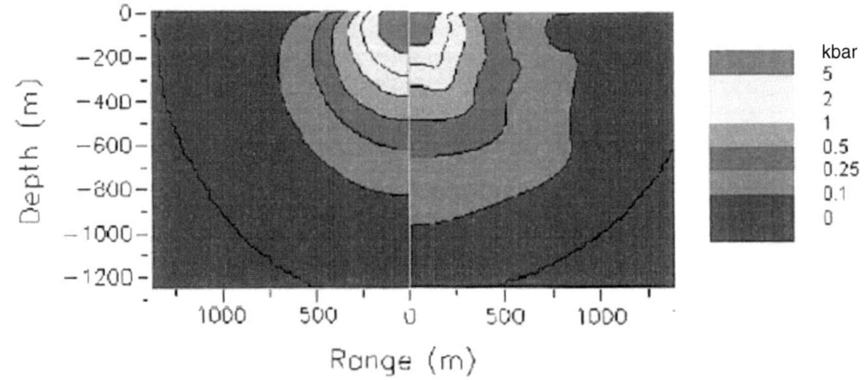

FIGURE 4.4 Peak stress contours for 300 kt earth-penetrator weapon (EPW) at 3 meters' depth of burst (left) and "damage equivalent" 5.6 Mt contact burst (right).

Uncertainty in range to effect and associated yield factors also arises from the damage (lethality) criterion selected. This is illustrated in Table 4.1, in which range to effect is shown as corresponding to various levels of strain, stress, and velocity for the 300 kiloton EPW at 3 meters' depth of burst and contact bursts of different yields. The upper section of Table 4.1 lists the range to effect (i.e., depth below the source) for different damage criteria and weapons. The columns labeled "Stress," "Strain," and "Velocity" list the down-axis range for those ground-shock properties associated with severe damage to a point target of the indicated hardness in a generic granitic rock geology for each of the listed weapons. Thus, for example, Table 4.1 indicates that a 1 kbar hard point target will be severely damaged at depths of 389 meters or less by a 300 kt EPW. If, however, the hardness of the target is only 0.25 kbar, then severe damage will occur at depths down to 501 m. The relative effectiveness of the other weapons is indicated by the other entries in the upper section of Table 4.1.

The information on range to effect for the 300 kt EPW and 5.6 Mt contact burst is obtained from Figure 4.3(b). As noted earlier, the near congruence of the two curves in Figure 4.3(b) indicates the equivalence of these weapons from a target damage perspective.

Similar information is provided in the columns of Table 4.1 labeled "Strain" and "Velocity." DTRA's improved assessment method characterizes the vulnerability of tunnel-like targets in terms of free-field strain or velocity normal to the tunnel surface rather than peak stress. The choice of damage criterion depends on target and weapon characteristics and is of no particular importance for the committee's considerations here. It notes, however, that a damage criterion of 0.2 percent strain is approximately equivalent to a peak stress criterion of 1 kbar, or a peak velocity criterion of about 10 m/s. The range of effect data in Table 4.1 for strain and velocity are obtained from Figures 4.3(a) (strain) and 4.3(c) (velocity).

The lower portion of Table 4.1, "Effectiveness Ratio," compares the relative effectiveness of the weapons listed in the upper portion of the table.

The effectiveness of one weapon in relation to another is indicated by the ratio of the range to effect for the two weapons. The yield factor based on hydrodynamic scaling is the cube of this ratio. As indicated in Table 4.1, the effectiveness of weapons with two target-damage potential equivalent (a 300 kiloton EPW at 3 meters' depth of burst and a 5.6 megaton contact burst) can differ by about 20 percent (yield factor variation of 60 percent), depending on the damage criterion selected.

TABLE 4.1 Range to Effect, R (in meters), and Its Correspondence to Stress, Strain, and Velocity for Different Weapons

	Damage Criterion							
	Stress			Strain		Velocity		
Yield/Source Location	0.25 kbar	0.75 kbar	1 kbar	0.15%	0.2%	5 m/s	10 m/s	15 m/s
Range (meters) to effect for								
300 kt EPW[a]	501	435	389	411	370	464	359	307
5.6 Mt contact	528	435	363	401	335	477	318	263
300 kt contact	182	150	125	138	118	164	111	94
1 Mt contact	275	226	186	208	175	247	166	140
	Effectiveness Ratio							
R (300 kt EPW) (m) / R (5.6 Mt contact) (m)	1.02	1.08	1.15	1.32	1.20	1.05	1.22	1.25
Yield factor	1.06	1.26	1.54	2.29	1.73	1.15	1.80	1.97
R (300 kt EPW) (m) / R (300 kt contact) (m)	2.75	2.90	3.11	2.88	3.14	2.82	3.23	3.27
Yield factor	21	24	30	24	31	22	34	35
R (300 kt EPW) (m) / R (1 Mt contact) (m)	1.82	1.92	2.09	1.98	2.11	1.88	2.16	2.19
Yield factor	6	7	9	8	9	7	10	11

[a]Earth-penetrator weapons at 3 meters' depth of burial.

TARGET DAMAGE PROBABILITY ESTIMATES

This section explores the destructive capabilities of various nuclear weapons—ones that are "contact burst" at the ground surface, ones that penetrate the surface, and ones that are burst in the air. The calculations were done using PDCALC (see Attachment 4.1 in this chapter), a tool used by DTRA and the United States Strategic Command (USSTRATCOM) to calculate the probability of severe target damage from nuclear weapons.

The HDBTs that are considered are those depicted in Figure 2.1. Also included are hard, surface point targets, such as missile silos. In Figures 4.5 through 4.7, the shaded areas delineate the range of target depth or target hardness of the vast majority of the 10,000 facilities that the Defense Intelligence Agency (DIA) believes to be targets, as discussed in Chapter 2. The vulnerability of deep underground targets is expressed in terms of DIA's GVNs (or, equivalently, peak stress). For surface and near-surface targets, hardness is expressed in terms of DIA's vulnerability numbers (VN or equivalently peak surface overpressure).

The deep underground C^3 complexes and missile tunnels are between 100 meters and 400 meters deep, with the majority less than 250 meters deep. A few are as deep as 500 meters or even 700 meters in competent granitic or limestone rock.

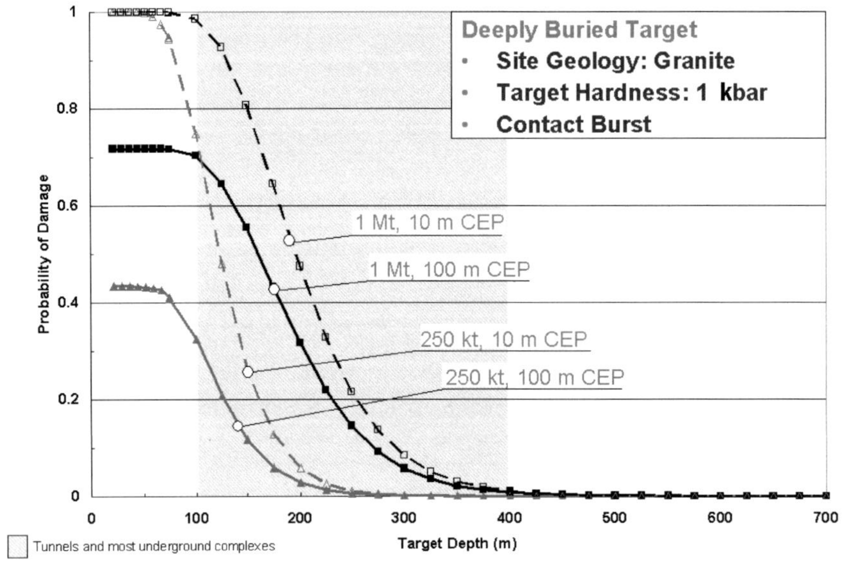

FIGURE 4.5 Effectiveness of contact bursts against some deeply buried targets. Note: CEP = circular error probable (i.e., accuracy).

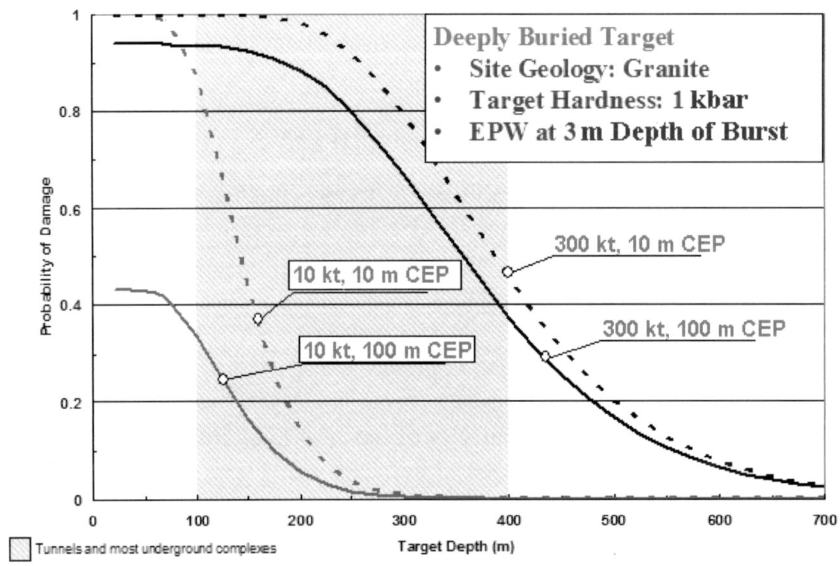

FIGURE 4.6 Earth-penetrator weapon (EPW) needs to be of sufficient yield to be effective against targets of interest. Note: CEP = circular error probable (i.e., accuracy).

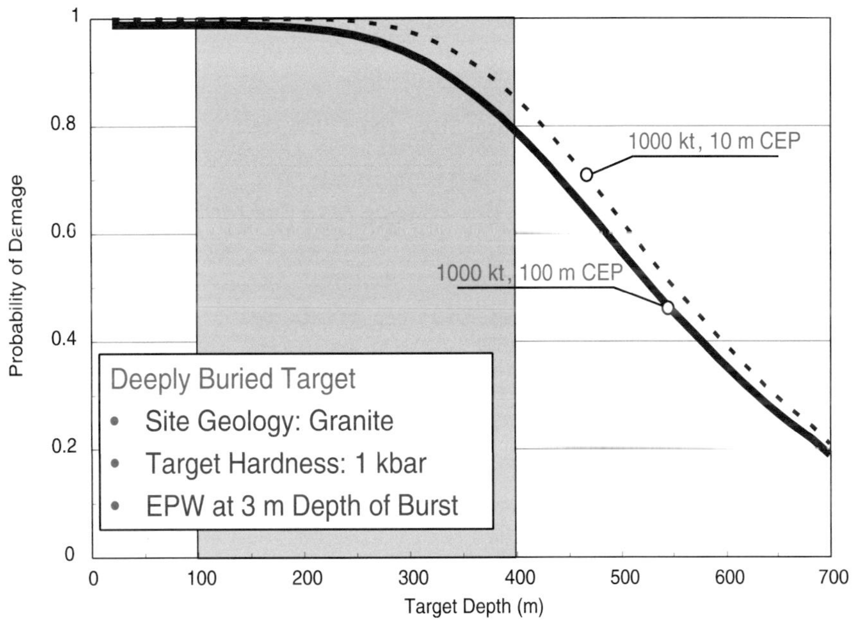

FIGURE 4.7 Earth-penetrator weapon (EPW) needs to be of sufficient yield to be effective against targets of interest. Note: CEP = circular error probable (i.e., accuracy).

Figures 4.5 through 4.7 are plots of the probability of (severe) damage of deeply buried targets in granite, as a function of the target depth. (For purposes of the calculations the targets are considered as points.) Figure 4.5 compares a 250 kiloton and a 1 megaton weapon, both of which are contact burst on the surface, for 100 meter and 10 meter CEP. Figures 4.6 and 4.7 are similar, but for 10 kiloton, 300 kiloton, and 1 megaton earth-penetrator weapons at 3 meters' depth of burst. Examining these figures, one observes the following:

- The effectiveness of a 250 kiloton contact burst is about the same as that of a 10 kiloton EPW, as expected from the analysis earlier in the chapter showing the 15 to 25 yield factor for equivalent ground shock.
- Accuracy (i.e., CEP) is a critical parameter, for contact weapons, and at low yields for penetrating weapons.
- For the target depths of interest, the most effective options examined are the 300 kiloton EPW and the 1 megaton EPW.

Included in Attachment 4.1 are additional figures and associated discussion on the influence of target hardness and CEP as well as information on nonsurface bursts.

Summary of Target Destruction

Following is a concise summary of target destruction:

- For deeply buried targets:
  —An EPW is more effective than a contact burst of the same yield. The probability of damage for a 300 kiloton EPW at 3 meters' DOB is equivalent to that for a 5 to 6 megaton surface-burst of the same accuracy.
  —For an EPW, yields in the range of several hundreds of kilotons to a megaton are needed to effectively hold deeply buried targets of interest at risk with a high probability of destruction.
- For surface and near-surface targets of interest, as shown in Figures 4.17 through 4.19 (in Attachment 4.2), detonating a weapon at its fallout-free height of burst could effectively destroy a target without producing local fallout, although significant casualties would result from the other weapon effects (as at Nagasaki and Hiroshima).

ATTACHMENT 4.1: PROBABILITY OF DAMAGE CALCULATOR

The Probability of Damage Calculator (PDCALC) calculates the probability of damage (PD) to targets caused by, for example, overpressure, dynamic pressure, cratering, and ground-shock coupling due to nuclear weapons effects. PDCALC can also calculate weapon radius or offset to a desired probability of damage. PDCALC handles a variety of targets, including soft urban/industrial buildings, shallow buried bunkers, bridges, silos, and deeply buried tunnels. This calculator is also used to make personnel casualty assessments.

Government agencies and contractors use PDCALC as an analytical tool for planning and studies regarding nuclear weapons effectiveness, weapon requirements, and target vulnerability and survivability. DTRA manages the development and maintenance of the calculator and sponsors the PDCALC Oversight Panel and the PDCALC Users' Group.

PDCALC is based on two Defense Intelligence Agency publications: *Physical Vulnerability Handbook—Nuclear Weapons*[11] and *Mathematical Background and Programming Aids for the Physical Vulnerability System for Nuclear Weapons.*[12] DIA provides the mathematical formulations embedded within PDCALC, as well as target vulnerability information used to develop target vulnerability numbers (referred to as VN). DTRA provides nuclear weapons effects data and algorithms and structural loading and response prediction methods required to develop estimates of vulnerability. While PDCALC does not contain any nuclear weapons effects models, it utilizes VN to specify the hardness of targets to various nuclear weapons effects such as overpressure, dynamic pressure, cratering, ground shock, thermal radiation, and initial nuclear radiation.

The vulnerability of a deeply buried target is given by a 10-character ground vulnerability number (GVN). GVNs are developed by DIA using the DUG1c ground-shock model. Results of depth-to-effect as a function of contact-burst yield are fit by a GVN, which is used by PDCALC. For heights of burst (HOBs) or depths of burst (DOBs) other than a contact burst, PDCALC makes use of a coupling curve to determine the equivalent contact-burst yield. This coupling curve was recently updated to incorporate the latest knowledge of energy coupling from near-surface nuclear explosions.

Currently under development is an improved GVN methodology in which DUGlc, the one-dimensional engineering ground-shock propagation code, is replaced by WinGS, a two-dimensional physics-based ground-shock code, as well as a finite-element-based tunnel response model. (See Chapter 5 for additional details.)

ATTACHMENT 4.2: INFLUENCE OF TARGET HARDNESS AND WEAPONS ACCURACY

Figures 4.5 through 4.7 assume the target hardness to be 1 kilobar. Figures 4.8 through 4.11 examine the sensitivity of the results of varying the target hardness from 0.75 kilobar to 1.5 kilobars. These figures show that for contact and buried bursts against deeply buried point targets, achieving a high probability of damage depends more strongly on improving accuracy (i.e., CEP) than on the target hardness, for the range of target hardness and yields of interest.

The scenarios for the targets examined in this study are likely to require the use of only one, or perhaps two, nuclear weapons, each of which should have an extremely high probability of damage. As discussed in Chapter 2 and earlier in this chapter, the HDBTs of interest are likely to have air defenses and to be protectively sited. Estimation of the probability, P_{sd}, that the weapon will survive through the defensive systems and detonate where desired (air, surface, or at depth) must consider such factors. This probability is not unity (i.e., P_{sd} will be less than 1). Since the calculated probability of damage, PD, assumes $P_{sd} = 1$, which is not realistic, the calculated probability of damage PD should be reduced by some factor of less than 1 for all cases. For an EPW, the PD should be reduced more than for a similar contact-burst or airburst case because the probability of penetration to the desired depth and the probability of successful detonation are additional factors to be considered; each of these probabilities is less than 1. Figures 4.12 and 4.13 examine how good the accuracy of a contact burst or EPW must be to achieve a very high PD against a deeply buried point target with a single weapon. (Again, P_{sd} is assumed to be 1.) Figure 4.12 demonstrates that very good accuracy—that is CEP ≤60 meters—is needed for a 1 megaton contact burst, and that targets of at most 125 meters' depth can be held at risk with a 0.95 PD. For an EPW, Figure 4.13 shows that a single 300 kiloton weapon eases the accuracy requirements to a CEP of 110 meters or less, with targets potentially as deep as 225 meters held at risk with a 0.95 PD.

Estimated hardness levels of the command, control, communications, and intelligence (C^3I) basement bunkers (Figures 4.14 through 4.18) are generally under 30 psi, the chemical weapon/biological

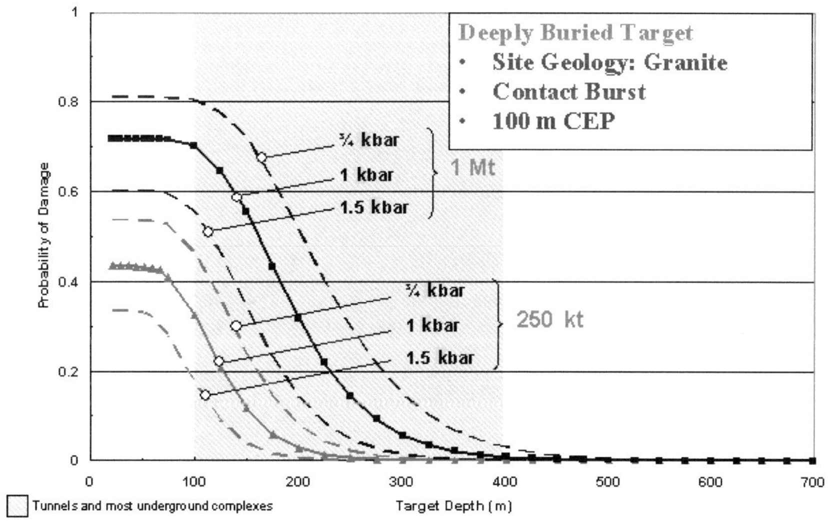

FIGURE 4.8 Contact burst with 100 meter circular error probable (CEP), or accuracy, against a deeply buried target. For a fixed CEP, effectiveness is not strongly dependent on target hardness.

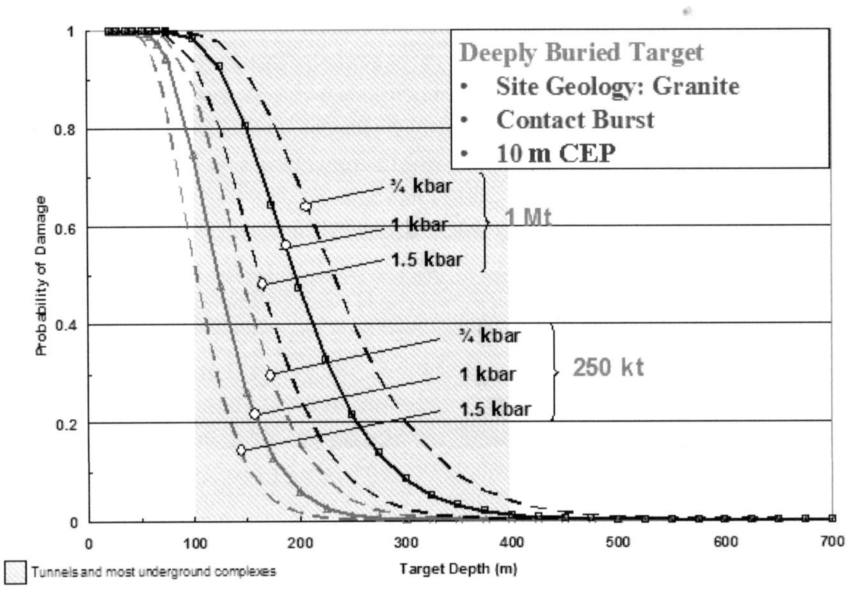

FIGURE 4.9 Contact burst with 10 meter circular error probable (CEP), or accuracy, against a deeply buried target. For a fixed CEP, effectiveness is not strongly dependent on target hardness.

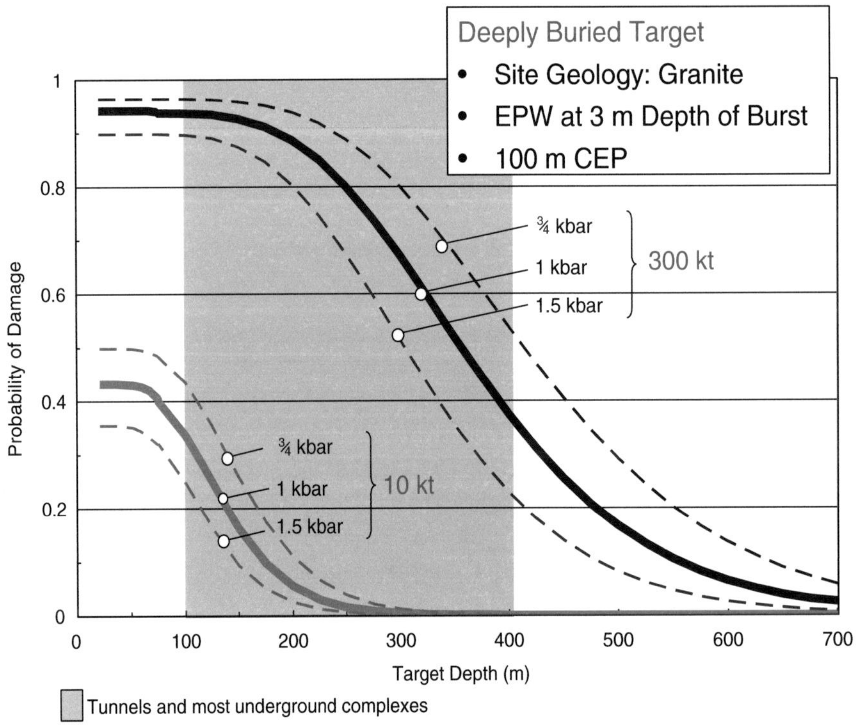

FIGURE 4.10 Earth-penetrator weapon (EPW) at 3 meters' depth of burst with 100 meter circular error probable (CEP), or accuracy, against a deeply buried target. For a fixed CEP, effectiveness is not strongly dependent on target hardness.

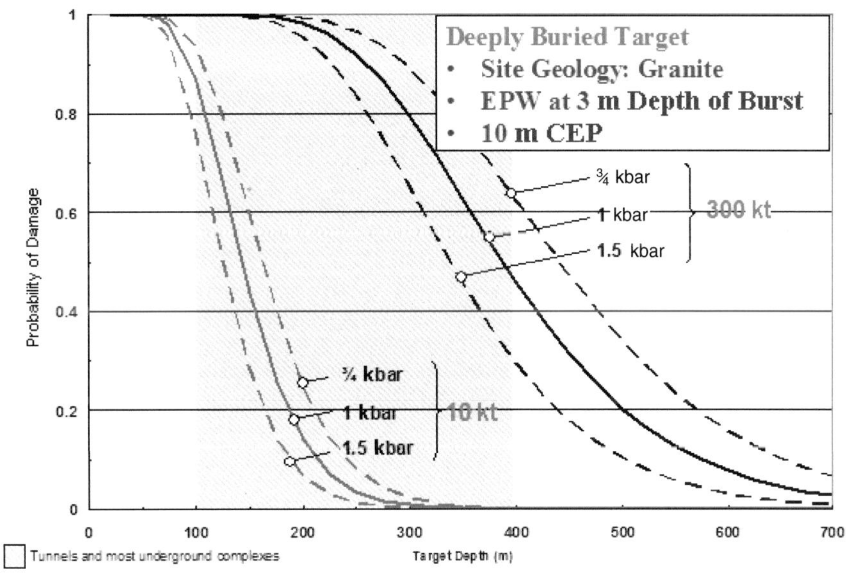

FIGURE 4.11 Earth-penetrator weapon (EPW) at 3 meters' depth of burst with 10 meter circular error probable (CEP), or accuracy, against deeply buried target. For a fixed CEP, effectiveness is not strongly dependent on target hardness.

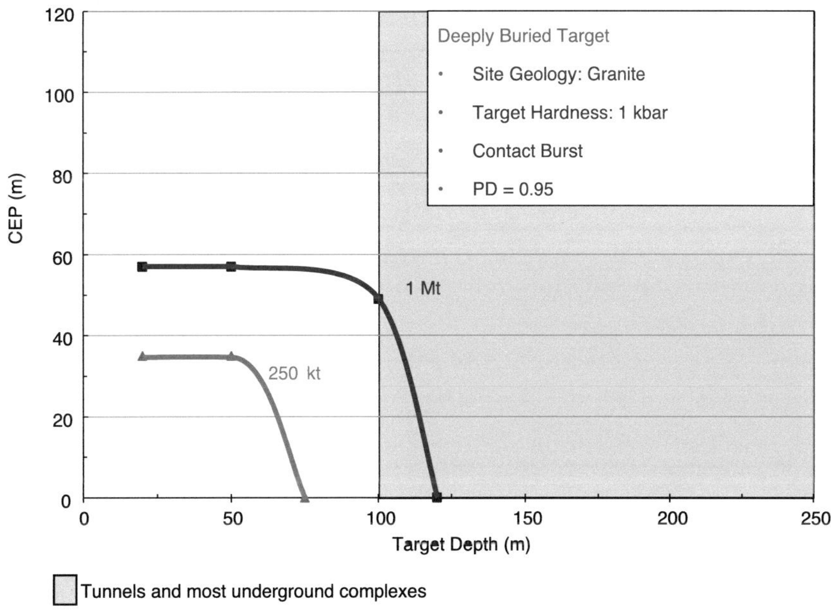

FIGURE 4.12 Contact burst against a deeply buried target. Very good accuracy and sufficient yield are required for a 0.95 probability of damage (PD) at limited target depth.

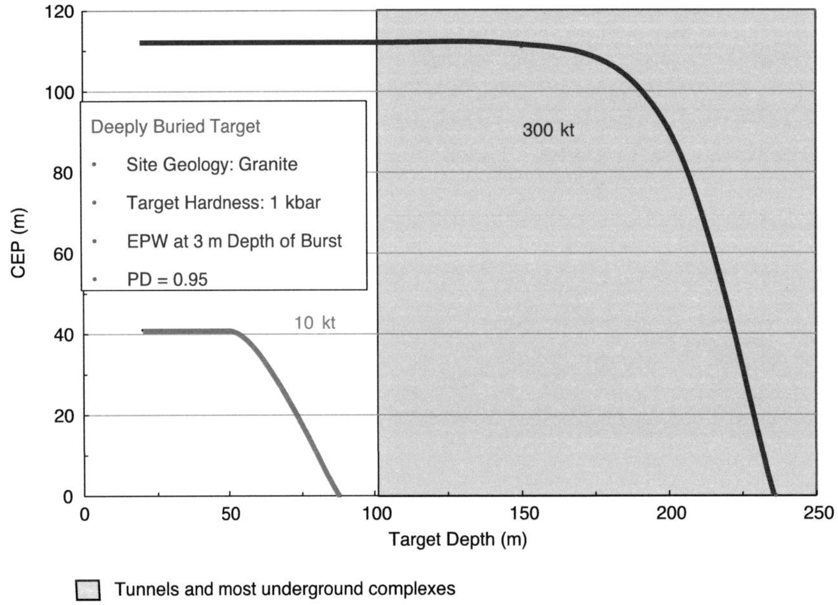

FIGURE 4.13 Earth-penetrator weapon (EPW) at 3 meters' depth of burst against a deeply buried target. Good accuracy and sufficient yield are required for a 0.95 probability of damage (PD).

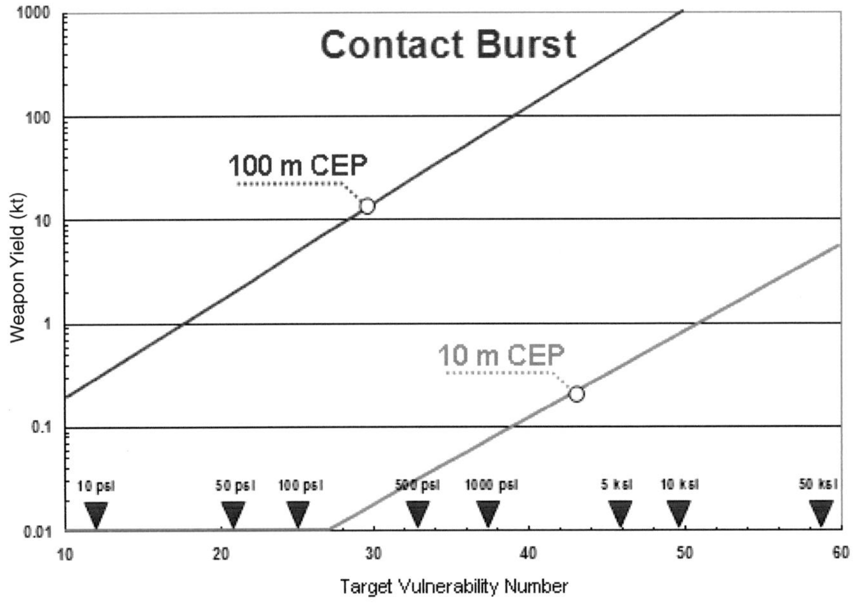

FIGURE 4.14 Contact burst against surface and near-surface point targets. Minimum yield to achieve a 0.95 probability of damage (PD) strongly depends on accuracy. Note: CEP = circular error probable; ksi = 1,000 psi.

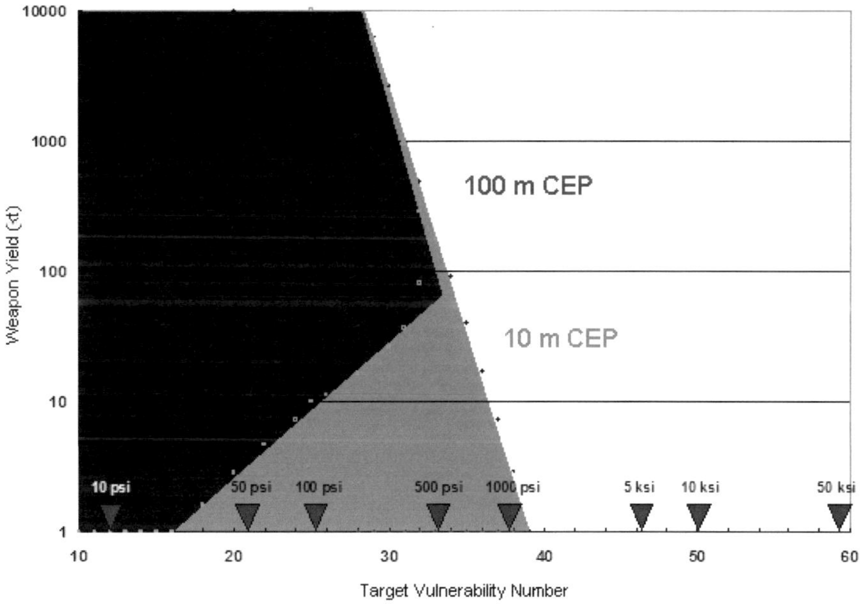

FIGURE 4.15 Fallout-free height of burst (HOB) 180 $W^{0.4}$: yield to achieve a probability of damage (PD) equal to or greater than 0.95 against surface/near-surface point target. Note: CEP = circular error probable; ksi = 1,000 psi.

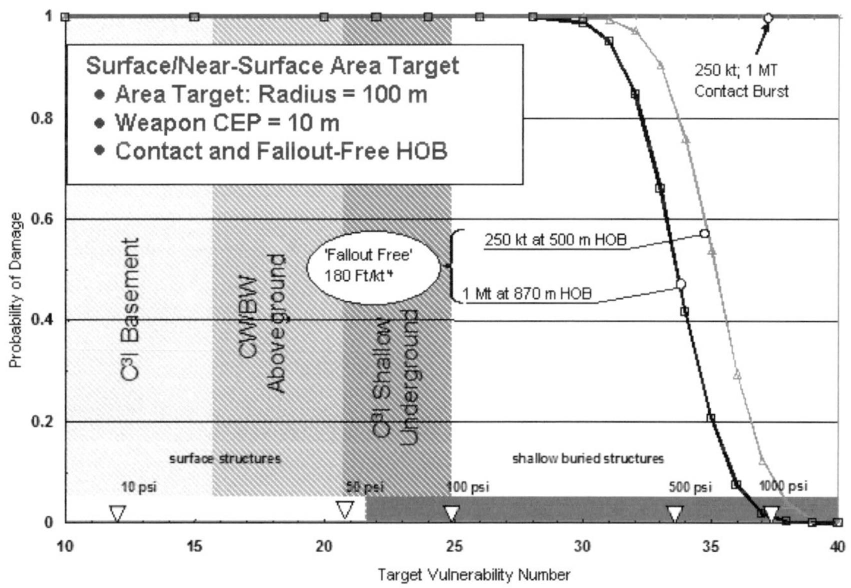

FIGURE 4.16 Large, hard, area targets vulnerable to contact or fallout-free height of burst (HOB), with weapon circular error probable (CEP) of 10 meters. Note: C³I = command, control, communications, and intelligence; CW/BW = chemical weapons/biological weapons.

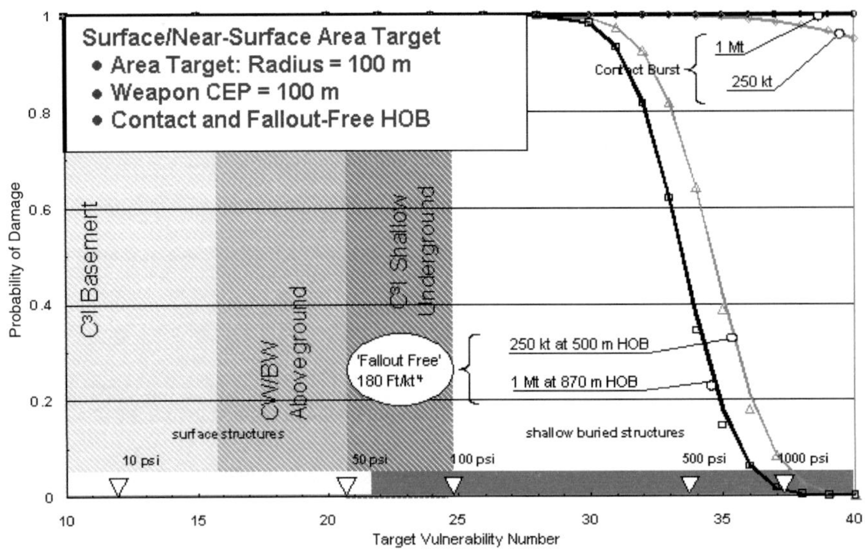

FIGURE 4.17 Large, area targets vulnerable to contact or fallout-free height of burst (HOB), with weapon circular error probable (CEP) of 100 meters.

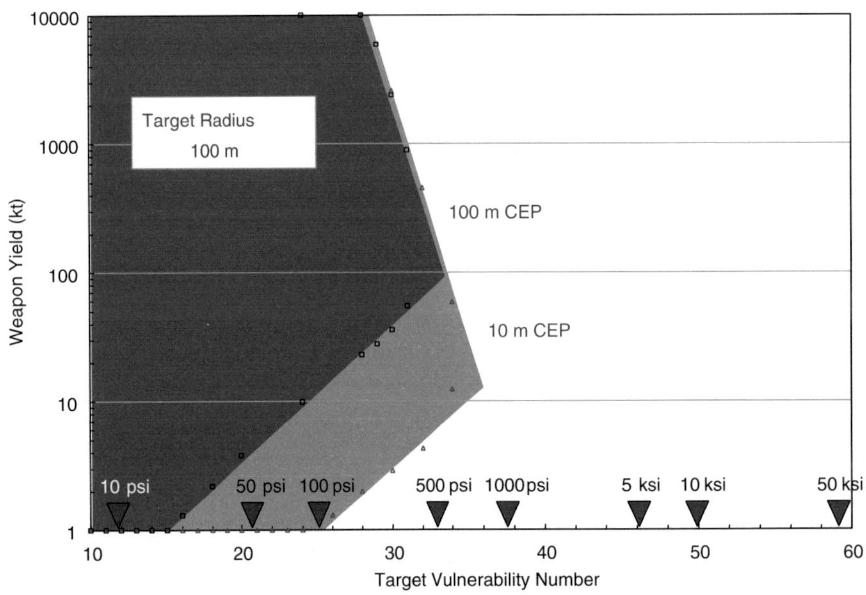

FIGURE 4.18 Fallout-free height of burst (HOB) 180 $W^{0.4}$: yield for probability of damage (PD) greater than 0.95 against surface/near-surface area targets. Note: ksi = 1,000 psi.

weapon (CW/BW) aboveground bunkers are on the order of 30 to 50 psi, and the shallow underground bunkers are 50 to 100 psi. Hardness levels of missile silos are generally in the range of 1,000 psi or less.

Figures 4.14 and 4.15 focus on surface and near-surface point targets. Presumably, EPWs would not be used for such targets, especially if other options were available that were effective and could ameliorate the collateral damage due to fallout. For these, this report concentrates on surface-burst and contact-burst weapons and the so-called fallout-free height of burst. The fallout-free HOB, as its name implies, is sufficiently high that the fireball produced by the nuclear explosion does not touch the ground. It therefore is not expected to generate significantly measurable local fallout, because no surface material is entrained, lofted, or dispersed. The fall-out free HOB increases with yield, W, as $W^{0.4}$.

Figure 4.14 shows that for surface and near-surface point targets, the minimum contact-burst yield needed to achieve a very high PD (i.e., 0.95) is strongly dependent on accuracy. For example, a yield of ~70 kilotons is required to achieve a PD of 0.95 against a 1,000 psi hard target if the CEP is 100 meters, whereas less than 1 kiloton is sufficient if the accuracy is 10 meters. Figure 4.15 is similar to Figure 4.14, except that the weapon is detonated at the fallout-free HOB.

Comparison of Figures 4.14 and 4.15 shows the following:

- The minimum yield for a particular target vulnerability level is somewhat lower for the contact burst owing to higher air-blast levels predicted at the lower detonation altitude.
- Since the lethality is proportional to $W^{1/3}$, for a given PD there is a finite limit to the hardness of the target that can be destroyed at the fallout-free HOB. In this case the maximum target hardness is ~600 psi (with ~70 kilotons) for a 100 meter CEP, and ~1,000 psi (with ~1 kiloton) for a 10 meter CEP. To attack harder targets with a 0.95 PD, a higher yield detonated at a height lower than its fallout-free HOB is required.
- The fallout-free-HOB constraint restricts the target set that can be held at risk with high probability. Further, increasing weapon yield or reducing CEP does not afford the same measurable increase in PD that may be achieved with a contact or very low airburst.

Figures 4.16 to 4.18 address hardened-surface and near-surface area targets that have a radius of ~100 meters. For these cases, an EPW is not a weapon of choice. Rather, weapons within two yields, 250 kilotons and 1 megaton, are compared at contact burst and at fallout-free HOB. Figure 4.16 assumes a 10 meter CEP, Figure 4.17 a 100 meter CEP.

The hardness levels of interest are shaded in Figures 4.16 and 4.17, with the most numerous targets in the ~5 psi to ~40 psi range. The figures show that for the target hardness range of greatest interest, any of the weapons considered will have a very high PD. The 250 kiloton weapon at its fallout-free HOB (i.e., 500 meters) is more effective than a 1 megaton weapon at its fallout-free HOB of 870 meters. This conclusion follows from cube root scaling (i.e., 250 kilotons at 500 meters gives the same peak overpressure as 1 megaton at 870 meters). Thus, the 1 megaton fallout-free HOB results in a correspondingly lower overpressure on the ground at comparable ranges. For the target hardness levels of most interest, the CEP is not especially relevant because of the extent of the target's size. The CEP will matter more as the area of the target shrinks.

Figure 4.18 is similar to Figure 4.15, except that the surface and near-surface targets are treated as area targets for this calculation. For targets of hardness ≤100 psi, a PD of at least 0.95 can be achieved with a weapon of less than a 10 kiloton yield and no fallout if the weapon is detonated at its fallout-free HOB. Similarly, a weapon with a yield of 10 kilotons to 100 kilotons, depending on the CEP, could destroy a 300 psi target with PD ≥0.95, with no fallout.

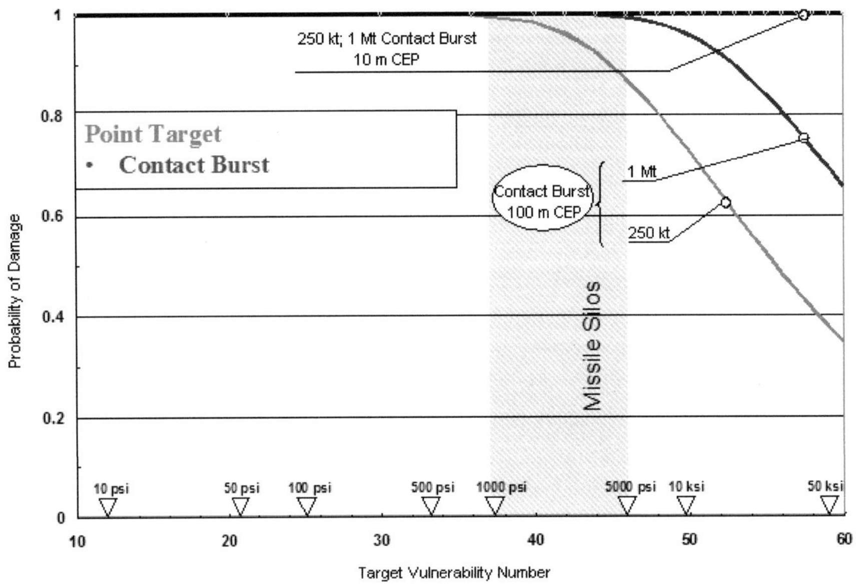

FIGURE 4.19 Accuracy and yield are important for hard point target destruction.

Figure 4.19 treats very hard point targets, such as missile silos. Only a contact burst would be effective. This figure shows the strong dependence of PD on accuracy and yield.

NOTES

1. Defense Nuclear Agency. 1991. *Effects Manual-1 (EM-1): Capabilities of Nuclear Weapons (U),* DNA-EM-1, Alexandria, Va., December (Classified).

2. Ground-shock-coupled energy includes the additional energy coupled to the ground by surface air blast for height of burst detonations.

3. The limiting equivalent yield factors are 99.4 and $1/f_{gs}(-0.05) = 49.9$ for yield and ground-shock-coupled energy, respectively.

4. Equivalence in terms of target damage criteria is discussed in the section entitled "Uncertainties in Weapon Effectiveness, Energy Coupling, and Ground-Shock Range to Effect."

5. Shel Schuster and Hal Zimmerman, 2003, "A White Paper on Energy Coupling from Nuclear and High Explosive Sources," Titan Systems Corp., San Diego, Calif., May 20. Also discussions with Dr. Zimmerman.

6. Based on a historical rule that requires a minimum depth of burst of 183 meters or a scaled depth of burst of 122 $m/kt^{1/3}$ to ensure containment at the Nevada Test Site for a carefully stemmed emplacement hole.

7. R.W. Terhune. 1978. *Analysis of Burial Depth Criteria for Containment,* UCRL 52395, Lawrence Livermore National Laboratory, Livermore, Calif., January.

8. The term "stemmed" means that the emplacement hole is very carefully backfilled in an attempt to prevent the release of radioactivity via the emplacement hole. It implies that multiple cemented plugs were inserted in addition to filling the emplacement hole completely from the device location to the ground surface.

9. Defense Nuclear Agency. 1991. *Effects Manual-1 (EM-1),* Chapter 3, "Cratering, Ejecta and Ground Shock," DNA-EM-1-CH-3, Alexandria, Va., December.

10. VN, vulnerability number, is a generic name given to a set of alpha numerics that describes the vulnerability of a target to a nuclear weapon. VNTK, a specific vulnerability number, is assigned by DIA to all facilities and personnel. DIA uses

dynamic and static pressure from a nuclear blast as the damage mechanism. GVN, ground vulnerability number, is a specific vulnerability number given to facilities below a specified depth. DIA uses ground shock as the damage mechanism.

11. Defense Intelligence Agency. 1992. *Physical Vulnerability Handbook—Nuclear Weapons,* Washington, D.C.

12. Defense Intelligence Agency. 1974. *Mathematical Background and Programming Aids for the Physical Vulnerability System for Nuclear Weapons,* Washington, D.C.

5

Fallout and Tools for Calculating Effects of Release of Hazardous Materials

NUCLEAR BURSTS AND FALLOUT OVERVIEW

All nuclear bursts produce radioactive fission debris in the form of gases that eventually condense into aerosols. But a burst near or under the ground surface will also entrain dust ejected from the associated crater or kicked up by the blast wave. Some of this dust will scour the fission debris and become radioactive. After being lofted by the rising fireball, the dust begins to fall out of the cloud, resulting in a fallout pattern on the ground. The size and intensity of the fallout pattern are heavily influenced by the weapon yield and height or depth of burst and by the meteorological conditions both at the surface and at altitude. There are several critical parameters in determining the fallout pattern: the yield of the weapon, the fraction of the yield derived from fission (versus fusion), the height or depth of burst, the size distribution of the entrained dust, the base surge or material lofted by ground motion (for explosions below the surface), and profiles of atmospheric conditions. The water and ice loading depend critically on the amount of ambient moisture in the atmosphere.

There are four heights of burst (HOBs)/depths of burst (DOBs) of interest for this study of nuclear weapons: (1) a surface, or contact, burst; (2) a low-altitude (below the fallout-free limit) airburst; (3) an airburst at the fallout-free height of burst; and (4) a subsurface burst.

The cloud from a 500 ton surface burst could rise to a few kilometers, whereas that from a 1 megaton burst would stabilize in the stratosphere with the top around 20 kilometers. A 500 ton surface burst would loft about 500 tons of dust that would be contaminated by the fission debris, whereas a 1 megaton burst would loft 300,000 tons. The amount of dust contaminated and lofted falls off rapidly as the height of burst is increased, primarily because above about 7 m/kt$^{1/3}$ there is no crater.

NOTE: The sections of this chapter addressing fallout are an amalgam of information from the following: *The Effects of Nuclear Weapons*, compiled and edited by Samuel Glasstone and Philip J. Dolan, 1977, U.S. Government Printing Office, Washington, D.C., §2.18–2.21, §2.23–2.31, §2.91–2.95, §2.99–2.100; *A Study of Base Surge Phenomenology,* Kaman Science Corporation, November 5, 1986; and *The DTRA Nuclear Weapons Effects Technology Information (NWETI) Fallout Guide*, 2004, by Joseph McGahan. They also contain material developed from discussions with members of the committee and presentations to the committee by David Bacon and Joseph McGahan, both of Science Applications International Corporation.

A burst at or above the fallout-free height of burst, as its name implies, produces aerosolized fission debris but no large particles, because the surface material does not mix with the remnants of the nuclear detonation. With high relative humidity, moisture in the cloud could form rain, ice, or snow that could scavenge the fission debris. Scavenging could also occur if the nuclear cloud encountered an ambient rain cloud. This height of burst increases with the yield of the weapon but is roughly 55 m/kt$^{0.4}$ (~870 meters for a 1 megaton burst). Even above this height of burst, there can still be fallout arising from a variety of mechanisms. The potential for long-term fallout on a global basis is well recognized from atmospheric tests, although dose-rate levels would be much lower than those from the ionized fallout from surface or subsurface bursts, with a potential impact on latent cancer rates.

A subsurface burst introduces additional processes for mixing of fission debris and dust. A main "mushroom" cloud like that from a surface burst is formed if the subsurface burst is not too deep (Figure 5.1); a low-level base surge is formed as the shock wave breaks through the surface (Figure 5.2). Thus, the subsurface burst may create two radioactive clouds that do not merge. As the depth of burst increases, the amount of activity vented to the atmosphere is reduced until the burst is completely contained, if the emplacement hole through which the device has been lowered has been carefully stemmed (70 meters for a 1 kiloton burst), which of course would not be the case for a weapon used to attack a target.

The primary short-term hazard from fallout is the external dose received from gamma rays. In addition, an internal dose can also be received through inhalation or ingestion of debris emitting beta and alpha radiation. Under most conditions, the external dose is the more dangerous, but the internal dose could have latent effects, specifically, cancer.

The total gamma ray activity in a measured fallout pattern is usually stated in terms of exposure rate in roentgens per hour (R/h) at 1 hour after the burst as if that activity were uniformly spread over an area of 1 square kilometer. For a 1 kiloton surface burst, this value is on the order of 9,000 R/h per square kilometer. Of course, the area of the fallout pattern is much larger than 1 square kilometer. The area covered by the associated dose that would cause at least a 50 percent probability of fatality is roughly 2.6 square kilometers per kiloton, assuming that people are in the open and exposed for just the first day after the burst. Over time the radioactivity decays, and eventually the fallout hazard decreases. Some radionuclides, such as cesium-137, have long half-lives, but most of the hazard is due to short-lived radionuclides. Within 1 to 24 hours after the burst, for example, the total gamma ray activity decreases by a factor of about 60.

The following sections focus on the problem of fallout associated with surface and subsurface bursts.

Nuclear Clouds

Nuclear clouds contain three classes of materials—radioactive materials, dust and pebbles, and water and ice. The first is by far the smallest contributor to the mass of the cloud, but it represents the greatest residual hazard and therefore is of prime importance. The radioactive materials consist of fission products from the weapon, as well as activation products produced by the interaction of the weapon's energetic radiation with the surrounding and/or nearby materials. When this radioactive material becomes attached to larger particles, it can result in fallout.

The dynamics process of the nuclear cloud starts with the ejection and then lofting of surface material. The cloud rises, entraining ambient air and moisture, cooling in the process. This cooling causes the materials to condense on nonvaporized residual material and causes these molten particles to coagulate into larger sizes. As the rising fireball and nuclear cloud continue to rise and entrain ambient

FIGURE 5.1 Example of a rising cloud, created by a shallow underground nuclear explosion at the Nevada Test Site.

air, at some point the fireball becomes cool enough that moisture begins to condense, releasing latent heat into the cloud. This makes the cloud more buoyant, leading to an increase in the updraft velocity, which leads to additional condensation, which in turn causes the agglomeration of wet particles. The turbulence of the cloud mixes the material composing the cloud. In addition, the large-scale circulation of the atmosphere drives the long-range transport of the cloud.

The larger particles in the nuclear cloud cannot remain in suspension, and hence they settle to the ground. For dry particles, this settling is called dry deposition. The process by which small particles are captured by precipitation that falls to the ground is called precipitation scavenging. Both can result in the surface deposition of radioactivity—that is, fallout.

FIGURE 5.2 Example of a base surge created by a shock wave breaking through the surface.

When caused by a low-airburst detonation, the dust mass loading in a nuclear cloud at about 10 minutes after the burst, when the cloud is stabilized, is roughly 0.1 Mt/Mt; the cloud from a surface burst has typically three times this amount of dust mass. The cloud consists of particles ranging in size from the very small ones (submicron to micron sized) produced by condensation as the fireball cools, to larger, surface material particles (10 to 500 micron sized) raised by the blast winds and the convectively driven afterwinds, to very large particles (up to millimeter sized) ejected from the crater. The radioactive cloud reaches a height of several miles before spreading out abruptly into a mushroom shape.

Dry Deposition and Precipitation Scavenging

Dry deposition and precipitation scavenging describe the processes resulting in the removal of radioactivity from the nuclear cloud and its deposition on the ground. Precipitation scavenging due to rainfall may produce "hot spots."

As mentioned earlier, fallout is a gradual phenomenon extending over a period of time. In the 15 megaton Bravo shot test at Bikini Atoll in 1954, for example, about 10 hours elapsed before the contaminated particles began to fall at the extremities of the 7,000 square mile contaminated area. The sizes of these particles ranged from that of fine sand (i.e., approximately 100 micrometers in diameter or smaller) in the more distant portions of the fallout area, to pieces about the size of a marble (i.e., roughly 1 centimeter in diameter and even larger) close to the burst point.

Particles in this size range arrive on the ground within 1 day after the explosion and will not have traveled too far (e.g., up to a few hundred miles) from the region of the burst, depending on the wind. Thus, the fallout pattern from particles of visible size is established within about 24 hours after the burst. This is referred to as early fallout, or also sometimes as local or close-in fallout. In addition, there is the deposition of very small particles, which descend very slowly over large areas of the ground surface. This is the delayed (or worldwide) fallout, to which residues from nuclear explosions of various types—air, high-altitude, surface, and shallow subsurface—may contribute. The size distribution of the fallout particles is critical to the determination of the level and geographical extent of the contamination due to fallout.

The details of the fallout are determined primarily by the radioactive materials inventory (which, in turn, is related to the device design and yield and the height or depth of burst), the amount of entrained mass, the injection height distribution, the particle size distribution, and the atmospheric transport—first locally and regionally, but ultimately including the global transport of residual radioactivity. Surface geology is critical to fallout. The prediction of fallout for shallow buried bursts is uncertain because, of the many underground tests performed by the United States, only three tests were at depths shallower than 20 scaled meters, and none was in rock. Therefore, confidence in predicting fallout from low-yield shallow and near-surface bursts in dry soil is higher than that for predicting fallout from such bursts in rock. Current targets of special interest are in mountainous terrain and urban environments. The yield of the tests at the Nevada Test Site were typically in the range of a few kilotons to a few tens of kilotons. While these shots were of less interest than when strategic yields were large (hundreds to thousands of kilotons), they are yields of interest for the current study.

The extent and nature of the fallout can vary widely. The actual situation is determined by a combination of circumstances associated with the energy spectrum, the yield and design of the weapon, the height of the explosion, the nature of the surface beneath the point of burst, and the meteorological conditions. In an airburst, for example, occurring at an appreciable distance above ground surface so that no large amounts of surface materials are sucked into the cloud (i.e., in the fallout-free HOB), the contaminated particles become widely dispersed. The magnitude of the hazard from fallout is thus far less than if the explosion were a surface burst. Thus, at Hiroshima (height of burst, 510 meters; yield, about 12.5 kilotons) and Nagasaki (height of burst, 500 meters; yield, about 22 kilotons), injuries due to fallout were completely absent.

On the other hand, a nuclear explosion occurring at or near the ground surface can result in severe contamination by the radioactive fallout. From the 15 megaton Bravo shot, the fallout caused substantial contamination over an area of more than 7,000 square miles. The contaminated region was roughly cigar-shaped and extended more than 20 miles upwind and more than 350 miles downwind. The width in the crosswind direction was variable, the maximum being greater than 60 miles. (It should be noted that the upwind contribution arises from the spreading of the nuclear cloud that occurs when the cloud hits the tropopause. In this extreme case, this process created a roughly circular cloud extending some 20 to 30 miles in all directions, owing to the fact that the convective velocities overshadowed the ambient wind speed.)

Time Domain of Physical Effects

There are roughly three phases in the generation of a radioactive cloud, with corresponding time intervals in which the fallout phenomena can occur:

- *The early time (less than 20 seconds).* In this first phase in the fireball, the mixing of ejecta and entrained sweep-up (dirt, vegetation, and rubble) with fission debris occurs.
- *The rise and stabilization phase (~10 seconds to 10 minutes).* In this second phase, the cloud rises and early fallout is produced. These effects are dependent on the local atmospheric profile, including temperature, relative humidity, and winds.
- *The late time (10 minutes to 2 days).* Wind transport and diffusion, as well as particle size, are major factors for the precipitation scavenging that occurs during this third phase.

Meteorology

The meteorological conditions that determine the shape, extent, and location of the fallout pattern from a nuclear explosion are the height of the tropopause, atmospheric winds, and the occurrence of precipitation. For a given explosion energy yield, type of burst, and tropopause height, the fallout pattern is affected mainly by the directions and speeds of the winds over the fallout area, from the ground surface to the top of the stabilized cloud, which may be as high as 100,000 feet. Furthermore, variations in the winds from the time of burst until the particles reach the ground perhaps several hours later—affect the fallout pattern following a nuclear explosion.

Shallow Underground Explosion Phenomena

When a nuclear weapon is exploded underground, a sphere of extremely hot, high-pressure gases, including vaporized weapon residues and rock, is formed. This effect is the equivalent of the fireball from an airburst or surface burst. The rapid expansion of the gas bubble initiates a ground shock wave that travels in all directions away from the burst point. When the upwardly directed shock (compression) wave reaches the surface, it is reflected back as a rarefaction (or tension) wave. If the tension exceeds the tensile strength of the surface material, the upper layers of the ground spall (i.e., split off into more-or-less horizontal layers), and these layers move upward.

When it is reflected back from the surface, the rarefaction wave travels into the ground toward the expanding gas sphere (or cavity) produced by the explosion. If the detonation is not at too great a depth, this wave may reach the top of the cavity while the cavity is still growing. The resistance of the ground to the upward growth of the cavity is thus decreased, and the cavity expands rapidly in the upward direction. The expanding gases and vapors can thus supply additional energy to the spalled layers so that their upward motion is sustained for a time or even increased.

As the ground surface moves, it continues to increase in height, and cracks form through which the cavity gases vent to the atmosphere. The mound then disintegrates completely, and the rock fragments are thrown upward and outward. Subsequently, much of the ejected material collapses and falls back, partly into the newly formed crater and partly onto its surrounding "lip." The material that immediately falls back into the crater is called the fallback, whereas that descending on the lip is called the ejecta. The size of the remaining (or "apparent") crater depends on the energy yield of the detonation and on the nature of the excavated medium. In general, for equivalent conditions, the volume of the crater is roughly proportional to the yield of the explosion. The material that is not fallback or ejecta becomes part of the base surge.

Part of the energy released by the weapon in a shallow underground explosion appears as an air-blast wave. The fraction of the energy imparted to the air in the form of a blast depends primarily on the depth of burst for the given total energy yield. The greater the depth of burst, the smaller, in general, is the proportion of shock energy that escapes into the air. For a sufficiently deep explosion, there is, of course, no blast wave.

Generation of a Base Surge

The base surge begins to form as the crater growth stops and entrained material in the column rising up to the main cloud begins to fall and expand radially along the ground surface. After the initial period of rapid radial growth, the base surge assumes a toroidal shape with a central depression. Throughout this period, additional material is propagated into the base surge from the smoke crown and central jet of the explosion (see Figure 5.2).

The base surge continues to grow vertically and radially along the ground surface until it reaches equilibrium with the surrounding air. Throughout the development of the base surge, its bulk density decreases as the surging flow expands and mixes with the surrounding air, while the entrained soil material settles to the ground surface. As the base surge reaches its stabilized geometry, it is relatively tenuous and remains airborne for a considerable period of time.

Experimental evidence indicates that the relative magnitude of the two stabilized clouds and the activity partition between these clouds is strongly dependent on the scaled depth of burst. The height of the stabilized main cloud depends on the initial energy transferred to the air and its mass loading. In general, however, the main cloud rises to an appreciable altitude, and it could present a serious radiological hazard at large distances from the burst point. The final distribution of the radioactive fallout from the main cloud depends on the features of the terrain and the local wind conditions at the time of detonation. Since the base surge cloud remains relatively close to the ground surface, the local winds are not expected to have a significant effect on the geometry of the base surge radioactivity. Therefore, a large fraction of the base surge cloud's residual activity is expected to fall within a relatively short distance of ground zero. For depths of burst of 2 to 3 scaled meters, the fraction of activity in the base surge is typically less than a few percent of the total activity.

CALCULATING THE EFFECTS OF RELEASE OF HAZARDOUS MATERIALS

The calculation of the effects of the release of a chemical or biological agent or of nuclear radioactivity can be broken down into several stages: (1) The first stage is the determination of a source term—specifically, how much agent or radioactive material is released at some initial time at some location. This requires estimating what is in the target being attacked, which in many cases is uncertain. The source could be a nuclear explosion or the intentional or accidental release of a chemical or biological agent. (2) In the case of a nuclear explosion, the next stage is the estimation of deaths and serious injuries due to prompt (immediate) air-blast effects, thermal effects, and initial nuclear radiation. (3) Radioactive material or chemical/biological agent released into the atmosphere is transported by winds until it either settles to the ground or is rained out. The transport also depends on the details of the terrain, and the material can be affected by temperature, radiation from the explosion, sunlight, atmospheric stability, and so on, during its transport. When the material eventually reaches the ground, it can be inhaled or absorbed into the skin of exposed individuals, and the total dose per individual can lead to adverse health effects (and death for large doses). The third stage of calculating effects thus addresses the potential for adverse health effects.

A number of computational tools have been assembled over the years to calculate the sequence of events listed above for nuclear explosions or other releases of radioactivity. Substantial efforts to validate the tools against experimental data have met with reasonably good success. With the heightened concern over chemical and biological agents during the past decade, many of the tools assembled for calculating the effects of nuclear radioactivity have been adapted to situations involving chemical and biological agents, and there has been an attempt to create a base of experimental data against which to validate the tools. Although this effort has met with some success, the results are much more fragmentary and uncertain than in the nuclear case.

In the following sections, the committee outlines the principal tools employed for the computer programs used to model the effects of the release of hazardous materials into the atmosphere and gives some assessment of the uncertainties involved at each stage of the calculation.

TOOLS TO ESTIMATE CASUALTIES AND OTHER EFFECTS

Two computer programs are in wide use to model the effects of the release of hazardous chemical, biological, radiological, and nuclear (CBRN) materials into the atmosphere and their collateral effects on civilian and military populations. The Hazard Prediction and Assessment Capability (HPAC) code was developed by the Defense Threat Reduction Agency (DTRA) and its predecessors to analyze nuclear, chemical, and biological releases for military studies and operational planning. Lawrence Livermore National Laboratory (LLNL) uses the NUKE code from Sandia National Laboratories to model prompt (immediate) effects from a nuclear explosion and the K-Division Defense Nuclear Agency Fallout Code (KDFOC) to analyze the spread of radioactivity. KDFOC was developed to model fallout from Plowshare tests, which involved nonweapon nuclear explosions designed to produce craters with minimal fallout. Parameters are adjusted to fit the available data from nuclear tests at the Nevada Test Site.

Operational Planning Tools of the Defense Threat Reduction Agency

The Department of Defense uses primarily the DTRA-developed HPAC for estimating effects resulting from a release of CBRN materials. HPAC predicts releases of hazardous material into the atmosphere, downwind transport, and the impact on civilian and military populations. HPAC is composed of a variety of software modules handling four principal functions: (1) generation of information on the source term such as type of hazardous agent, agent mass, release rate, and geometry; (2) weather prediction and data retrieval; (3) determination of source term transport and dispersion code based on the Second-Order Closure Integrated Puff (SCIPUFF) model; and (4) prediction/indication of human effects and casualties. HPAC was designed to be fast running and to be used on a laptop computer with no connection to outside resources.

Using a stochastic simulation process, HPAC and KDFOC predict the dispersal and effects of CBRN hazardous materials based on three general tool components: for hazard sources, weather and transport, and effects. The source term component estimates the initial cloud dimensions, particle size distribution, and radiation attachment. The transport and dispersion component uses historical, observed, or forecast weather data to estimate the dose rate. Finally, the effects component integrates the dose rate to create hazard plots, casualty tables, population distribution, and prompt-effect circles for nuclear incidents.

Attachment 5.1 in this chapter provides descriptions of the following elements of the DTRA-developed codes and data:

- Source term codes
  —Chemical and biological sources
  —Liquid sources
  —Nuclear and radiological sources
- Transport and dispersion codes
  —HPAC SCIPUFF model
- Personnel, health, and environmental effects codes
- Databases
  —Population data
  —Agent fate
  —Weather
  —Terrain

Tools Used by Lawrence Livermore National Laboratory

The Lawrence Livermore National Laboratory has been engaged in modeling the effects of nuclear explosions since the 1960s. As with the HPAC code described above, many of the tools are under continuing development, motivated in part by advances in computer power that allow more complex calculations to be performed in a short time. In addition, the recent need to estimate the effects of a chemical or biological incident has introduced new phenomena that have to be included in the models.

The calculations requested for this study were performed with KDFOC4, the 1990s version of the LLNL fallout model. KDFOC was originally written to model fallout from Plowshare tests, which involved nuclear explosives designed to produce craters with minimum fallout. Consequently, the code's emphasis has been on surface and buried explosives, and parameters have been adjusted to fit the data available from nuclear tests at the Nevada Test Site and the Pacific Proving Ground.

Following an explosion, KDFOC4 defines an effective nuclear mushroom cloud, characterized by distribution of activity and particles as a function of particle size and altitude. The computer model accepts a description of the wind field and then slices the effective cloud into overlapping disks. Each disk characterizes the cloud's initial altitude, a particle size, and associated radioactivity. The horizontal motion of the disks is governed by the wind direction; vertical motion depends on the particle size, and each disk grows in size according to a scale-dependent diffusion. Each of these disks is blown by the wind field and eventually falls to the ground at its settling velocity. When all of the disks have reached the ground, the total radioactivity can be found by adding the contribution of all of the disks.

About 10 surface and belowground bursts at the Nevada Test Site produced well-characterized fallout patterns. Following a detailed study by DTRA (then the Defense Nuclear Agency), a factor-of-two agreement was found in comparing KDFOC predictions with the measured radiation contours. Another comparison with the 15 megaton Bravo test (part of Operation Castle) produced comparably good agreement with the model calculations.

KDFOC4 allows variation in the source term, the wind field, the degree to which the population is sheltered, and the terrain (although for this committee's study the terrain was assumed to be flat). The nuclear device is described by a fission/fusion ratio with equivalent fission yield based on a mix of Plowshare and laboratory experimental data. For a systems study, the wind field is described in 16 sectors of 22.5 degrees, and the fatalities are averaged separately for each of the wind directions. Expected fatalities are calculated using historical data on the wind field. The modeled population can be out in the open, partially sheltered, or completely inside.

Fatalities are computed as resulting from prompt effects and from local fallout. As with HPAC, the population data are taken from the LandScan 2002 database maintained by the Oak Ridge National Laboratory.[1] Prompt effects due to the initial heat, blast, and radiation are computed with the NUKE code developed by Sandia National Laboratories. The acute-fallout biological effects are based on the model in Glasstone and Dolan's *The Effects of Nuclear Weapons*, which associates a probability of lethality with a given dose of radiation.[2] The LD^{50} (i.e., the level of radiation at which 50 percent of the population would die) for these calculations is set at 4.0 sieverts (400 rems). Acute fatalities are defined as those that occur within the first 60 days after the explosion; deaths from latent effects of fallout are those that occur subsequently. However, no attempt is made to account for global fallout or for low radiation doses (less than 1 millisievert [100 millirems]).

SOURCES OF UNCERTAINTY AFFECTING MODEL PREDICTIONS

It is very difficult to give an overall assessment of the accuracy of the models described in this chapter. In this section the committee summarizes the state of the art in each of the major modeling areas and then brings these together to evaluate the overall uncertainty.

Source Term

The source term for nuclear explosions is a fit to the 10 or so surface-burst and buried-burst tests that were carried out at the Nevada Test Site in the 1960s and 1970s. It has also been tested against more limited data from the Pacific Proving Ground and from unintentional releases of radioactivity (e.g., from tests that vented) that then produced fallout some distance from the initial explosion. These tests were primarily integral experiments that measured radiation on the ground after the events. Most were done under benign meterological conditions, with low winds and relatively flat terrain. At a workshop held by DTRA (then the Defense Nuclear Agency), it was determined that it was possible to model all of the data to within about a factor of two.[3]

The source term for chemical and biological agents released after a conventional or nuclear attack on a facility is derived from a much sparser set of data. DTRA has carried out an extensive testing program to attempt to simulate such events, but the testing suffers from two factors not present in the nuclear case. First, the lack of a "signature" for tracing particles makes it much more difficult to detect the ultimate fate of the chemical and biological agents released. Second, variations in the effects of an attack with conventional weapons depend so strongly on such variables as how close the explosion is to the containers; whether the agent is liquid, solid, or gas; how well the explosive penetrates the target area; and so on that it is virtually impossible to predict, for likely real-world situations, the amount of agent lofted to the atmosphere for transport out of the immediate area. Third, the ultimate effects of a biological plume can be influenced by the inherent stability of the particular agent targeted (spores, vegetative forms, virions, and so on) and probably by whether the material is wet or dry. Nonetheless, it is probably possible to bound the source term through use of the DTRA experimental database and models, although this committee has not seen the analysis and results to make those judgments. Consequently, this study takes the position that a parametric analysis with an assumed fraction of released or lofted agent would be the best way to model the range of possible outcomes in our knowledge of the event.

DTRA has not provided a quantitative evaluation of its model for predicting chemical/biological agent release as tested against the results of its experiments, but it has suggested that lofting of between 0.1 and 5 percent of the stored agent is a plausible number for release of "transportable" agent,[4] and it proposed a bimodal distribution in particle size as a way of characterizing the releases.[5]

Transport and Dispersion

Once fallout or chemical or biological agent (specified in the source term as the amount of particles of each size as a function of altitude) is dispersed, then winds carry the particles away from their initial location, they spread horizontally owing to normal diffusion, and they settle to the ground under the force of gravity. When all of these processes are modeled, the results combine to produce contours of radiation or agent concentration at distances from the initial source.

Quite sophisticated transport models are used, and these are likely to be fairly accurate when the meteorology is well defined—that is, when there are well-specified wind directions, small or moderate turbulence, no rain, and good knowledge of the topography. All of the computer models are planned to include provisions for these factors (e.g., rain), but there has been very little testing of how well the models work in complex topography or stormy conditions. It is probably most important to incorporate major topographic features such as mountains into the models, since these will always have a major effect on the results compared with what a "flat" model would predict for flat terrain in such regions.

The validation of models for particle transport is intimately connected to knowledge of the source term, and in most cases it is hard to disentangle the two factors. However, as described in the section above, the combined results produce about a factor-of-two agreement for the available nuclear test data. DTRA has also modeled the Russian release of anthrax at Sverdlovsk (and is able to produce reasonable agreement with the published results).[6] Finally, over the years LLNL has used even more sophisticated models of atmospheric transport to characterize accidental releases of nuclear, chemical, or biological materials and has achieved some success in predicting global transport of such substances (e.g., from the accident at Chernobyl).

The existing state of the art in tracking and predicting atmospheric transport and dispersion of hazardous material was reviewed in another National Research Council study,[7] whose recommendations should help guide future work in this area. Although planning the needs of the military and the emergency response community are somewhat different, the commonalities are also very large. These two groups should thus make special efforts to compare and utilize the results of one another's research.

For modeling particle transport to predict the effects of hazardous materials dispersed in nuclear attacks on hardened targets or attacks with conventional weapons on chemical and biological agent facilities, it is probably more important to include more factors like topography in existing models than to build more sophisticated transport models. The exact distance at which hazardous materials of a particular concentration are deposited may be incorrectly estimated, but unless deposition is near the boundary of a populated area, that assumption will often not affect the predictions of casualties (the casualties will just occur in different places).

Additional Factors

The strength and moisture content of soil are known to have significant effects on the size of the crater produced by a nuclear explosion at a given depth of burst, but the specific relationship is uncertain. The relationship between the amount of energy trapped in a nuclear cloud and the crater size is not the same for all soil materials. Some materials are inherently stronger than others (granite as opposed to sand), but it is not clear that soil strength is more significant than the total energy required to move the weight of the soil and form a crater. Although it has not been proven, there is evidence that the most significant factor related to fallout is the efficiency of energy coupling. That is, for some soils, less energy is required to move a given mass of soil than for others. It is further believed that the key soil ingredient that improves this efficiency is moisture content. Water is a soil constituent that expands more per given energy input than do most other soil constituents.

A more complex aspect of prediction of the effects of fallout is the size distribution of soil particles in the nuclear cloud. Some of these particles may be picked up from the ground surface by the turbulent flow and not be radioactive. The extent of pickup will depend primarily on soil type, although the complete nature of the dependence is not known.

Weapon placement and environmental material around the burst also have an impact on amounts of fallout produced. Experiments are needed to quantify this impact. Weapon design and yield also affect the amount of fallout produced. In addition to using lower yields in earth-penetrator weapons (EPWs), reducing the amount of fission energy in the design of EPWs would also reduce fallout.

Another important uncertainty factor affecting the amount and impact of fallout is fractionation, a process whereby fission products become attached to the soil particles in the cloud stem and fireball. Both the base surge and ejecta—including radioactive material continuum that breaks up into discrete particles involving fission products—are strongly dependent on surface geology in the area of the explosion.

Casualties

Since the earliest days of the nuclear age, the nuclear radiation community has built up a detailed methodology for correlating radiation dose with the probability of death or serious injury. All of the nuclear test models capture this information. One source of uncertainty in the models concerns the determination of LD^{50} (the dose at which 50 percent of the population will die) and the long-term impacts of low-level radiation. The overall uncertainties in estimating casualties are in the tens of percent range and are smaller than most of the other sources of uncertainty in the models.

A second factor source of uncertainty is the amount and concentration of radiation present shortly after a nuclear explosion and its decay with time. To calculate this precisely requires knowledge of the details of a subsurface nuclear explosive (e.g., the amount of fission versus fusion) and how much activation occurs in different soil compositions. The models usually employ relatively simple algorithms for these effects and typically assume that the initial radioactivity decays with time to the negative 1.2 or negative 1.3 power. It is worth improving these prescriptions, but the errors are in the tens of percent range.

The chemical and biological communities have begun to focus on the real-world potential lethality of various toxic chemical and biological agents, but the variation in lethality, agent route exposure, agent lifetime, sensitivity to environmental factors, and releasability in virulent form is vastly wider than that for nuclear fallout. Current models incorporate these factors to some extent, but no sensitivity studies were presented to the committee that describe the ways in which each of these factors affects the lethality of various agents (and nearly every issue reduces the potency of an agent). It is strongly urged that such sensitivity studies be done and reviewed by appropriate scientific and medical experts and emergency responders, who are equally important in resolving these questions.

Overall Uncertainty

Careless use of the models to predict casualties from the use of nuclear EPWs can lead to estimates that are off by more than a factor of 10, and by much more than that when chemical and biological agents are targeted (and usually on the high side in these cases). Intelligent use of models to predict the effects of nuclear events can reduce this uncertainty to a factor of 3 if the wind direction, percentage of a population likely to be sheltered, and other meteorological and topographic factors are known and taken into account. Although the number of casualties at any precise location may not be accurately modeled, there is much less uncertainty in estimates of total casualties for populated urban areas.

The effects of release of chemical and biological agents are much more uncertain. However, a rule of thumb might be to use the code as well as possible by including all relevant phenomena, and then bounding the problem by a factor of a few in either direction, recognizing that casualties might in some cases be much less if the weapon failed to work or impacted off target or if the agent were particularly susceptible to environmental degradation. Studies of the conditions of release (e.g., whether from attacks occurring at night versus during the day) can also help elucidate the sensitivity of the results to environmental and other factors.

SUGGESTIONS FOR IMPROVING TOOLS

Based on its examination of the codes discussed in this study, the committee believes that tools such as HPAC and KDFOC (and others), should be upgraded to support more precise estimates of the health and environmental consequences of attacks on hard and deeply buried targets. In particular, this upgrading should include the following:

- Improve the characterization and validation of the nuclear source terms for a more detailed set of the geological and meteorological conditions characterizing the locations of hard and deeply buried targets;
- Incorporate global effects, including environmental impacts (on crop production, fishing, and so on) and low-dose-rate effects on humans across the full demographic spectrum, from infants to the elderly, using results from Chernobyl as well as from the Nevada Test Site and Pacific Proving Grounds;
- Address the needs of emergency responders;
- Refine population databases to include demographics and to account for movement of people, migration, and evacuation;
- Develop, integrate, and maintain three-dimensional urban and topographic databases;
- Validate the transport models more thoroughly, including over a broader range of realistic environmental conditions (e.g., topography, microclimates, and so on); and
- Substantially improve the chemical and biological models in the HPAC code and other tools as follows:
  —Document and expand the verification and validation of the source term for a wider variety of scenarios, and
  —Determine lethality and other parameters for each relevant chemical and biological agent from a consensus of the relevant technical communities and document them. In particular, describe the wide variability in human susceptibility to biological agents and the consequent uncertainty in predicting the effects of their dispersal as the result of an attack on facilities for the storage and/or production of agents.

The committee also suggests that it would be useful for DTRA to convene periodic workshops to compare the results of all widely used predictive tools against experimental data and sets of benchmark problems and to publish the results.

ATTACHMENT 5.1: ELEMENTS OF DEFENSE THREAT REDUCTION AGENCY CODES AND DATA

Source Term Codes

Mathematical representations of physical processes that produce a hazardous source are called source term models. Source terms define the characteristics of materials initially released in an explosion for use in modeling downwind transport and dispersion. Examples of HPAC source terms include explosions, sprays, and airborne releases (dumping) of bulk powders or liquids. Each of these means of agent dispersal can be described mathematically by equations derived from the laws of physics or by empirical correlations determined through experimental observations.

Chemical and Biological Sources

HPAC includes complex methodology to calculate source terms for breached containers in a chemical/biological facility, and models are being developed for incidents at industrial facilities. The initial versions of HPAC, prior to HPAC 4.04, use a table lookup system for predicting releases from breached containers. The user states the type and amount of agent that is contained and estimates a general damage category (light, moderate, severe, total). HPAC associates the percentage of agent released with each damage category. For biological agents, 10 percent, 20 percent, 30 percent, and 40 percent of the agents are released under light, moderate, severe, and total damage, respectively. For chemical or industrial agents, the corresponding percentages of agents released are 15 percent, 30 percent, 45 percent, and 60 percent. Note that the committee views these large percentages of released agent as unrealistic and concurs with DTRA that 0.1 percent to 5 percent is plausible. For HPAC 4.04, the legacy table lookup system is no longer used. Instead, for a user-selected damage category, the estimated release is physics-based, using three modeling steps: damage state, container release, and expulsion/cloud. The overall process is to determine the damage to the facility (containers and building), to predict the container releases, and to calculate the source term parameters for the SCIPUFF atmospheric transport and dispersion model.

Liquid Chemicals Source

Bulk liquid dissemination is the dumping of some amount of bulk liquid agent from an aerial platform. The aerodynamic forces experienced by the bulk liquid in the air lead to the formation of initial droplets, followed by further evolution of the distribution of droplet size owing to liquid evaporation and aerodynamic breakup.

A pool of liquid agent on the ground may also be a source of airborne hazards. Liquid agent may be spilled from a container or munition and then evaporate, forming vapor clouds that are transported downwind.

Pool evaporation constitutes a continuous source of airborne hazardous material over a period of time. Parameters describing a pool evaporation source are the rate and duration of evaporation and the initial cloud sizes generated. The number and size of clouds or "puffs" produced reflect the dimensions of the evaporating pool. The duration of the release is dependent on the initial depth of the pool and the rate of evaporation.

Nuclear and Radiological Sources

Radiological material, fission fragments, and nuclear radiation can occur from multiple releases or events, which include the following:

- Explosive release of radiological materials to energetically and quickly disperse the material (e.g., use of a relatively small amount of explosive to scatter the material in the air or over a surface);
- A nuclear reactor accident (or deliberate unauthorized action) during which reactor fuel and spent-fuel particles are disseminated into the atmosphere and/or water; and
- Detonation of a nuclear weapon or device on or near any target.

The following adverse environments would be created from detonation of a nuclear weapon or improvised nuclear device:

- Blast and shock,
- Large thermal output,
- Electromagnetic pulse,
- Initial nuclear radiation (primarily gamma rays and neutrons), and
- Residual radiation.

Some of these effects also can be present when an adversary deliberately disseminates radiological material.

Factors of concern when modeling dispersal of nuclear radiation include nuclear yield (if a detonation is involved), fallout/rainout, particle size, wind shear, geology, burst (or dispersal) height, type of ionizing radiation, prevailing winds and/or water currents, and migration, half-life, and human bioeffects.

HPAC uses two source term models for fallout: K-Division Defense Nuclear Agency Fallout Code (KDFOC) and New Fallout Code (NEWFALL). These models use the weapon parameters of yield, height or depth of burst, time, and fission fraction (fraction of nuclear energy produced by fission) to estimate the weapon output. KDFOC was jointly developed by LLNL and the Defense Nuclear Agency in the 1960s and is used for buried bursts. It is an empirical code based on tests at the Nevada Test Site that define an effective nuclear mushroom cloud, characterizing its initial altitude, particle size, and associated quantity of radioactivity. NEWFALL, used for shallow buried bursts, surface bursts, and airbursts, models particle size, agglomeration, and radiation distributions. NEWFALL is substantially based on the legacy Defense Land Fallout Interpretative Code (DELFIC) and on both British and Nevada Test Site (NTS) tests.

What Are the Major Limitations or Uncertainties?

Within the methodology, the level of uncertainty in HPAC code source term is a significant concern. Current HPAC capabilities for predicting nuclear effects were designed for use in the Cold War and massive strike scenarios. For these considerations, HPAC is more than accurate enough. Additionally, HPAC's fallout particle size and radionuclide inventories match high-end estimations for most nuclear tests. All codes for fallout are sensitive to weapon yield, height of burst, and target environment. No code takes into account the uncertainty associated with weapon yield or HOB. The current codes also do not explicitly account for secondary activation from the target-area environment. KDFOC accounts for secondary activation in the base for buried bursts, but only within the limits of the test data available. NEWFALL does not account for any secondary activation for shallow buried bursts, surface bursts, and airbursts. Depending on the target environment, the current codes may underpredict the associated fallout by up to several orders of magnitude. Assuming that the target area is comparable in geology and environment to the Nevada Test Site, the model source term for fallout, independent of weather, is well within an order of magnitude of error. Work is underway to more accurately estimate the effects of a single weapon at much higher fidelity, but this feature will not be operational for several years.

Transport and Dispersion Codes

Background

Atmospheric transport and dispersion modeling is one of the most essential functions of hazard assessment codes, largely because these processes play a major role in the evolution of particle and agent concentration profiles in the atmosphere. Most such models use the advection-diffusion equation, a solution that gives the concentration profile for a cloud as a function of time and location. HPAC uses

SCIPUFF to estimate the dose-rate matrix. SCIPUFF is statistically based, with a three-dimensional wind field involving shear owing to terrain interaction.

HPAC Second-Order Closure Integrated Puff Model

SCIPUFF is a Lagrangian model for the transport and atmospheric dispersion of material. The numerical technique employed to solve the dispersion model equations is the Gaussian puff method, in which a collection of three-dimensional puffs is used to represent an arbitrary time-dependent concentration field. The turbulent diffusion parameterization used in SCIPUFF is based on the second-order turbulence closure theories of Donaldson[8] and Lewellen,[9] providing a direct connection between measured velocity and the predicted dispersion rates.

The second-order closure model also provides the probabilistic feature of SCIPUFF through the prediction of the fluctuation in particle concentration. In addition to giving a mean value for the concentration, SCIPUFF provides a quantitative value for the random variation in concentration due to the stochastic nature of turbulent diffusion. This estimate of uncertainty is used to generate a probabilistic description of dispersion and gives a quantitative characterization of the reliability of the prediction. For many calculations of dispersion, the prediction is inherently uncertain owing to a lack of detailed knowledge of the wind field, and a probabilistic description is the only meaningful approach.

What Are the Major Limitations or Uncertainties?

The inputs for wind and weather will involve significant uncertainties. HPAC uses weather files (or flow fields) generated by others—the Air Force Weather Service, the Navy Fleet Numerical Operations Center, applications of the DTRA Operational Multiscale Environment Model with Grid Adaptivity (OMEGA), the National Oceanic and Atmospheric Administration, and allied weather models—or it operates on simple input such as a single wind vector, uniform throughout the domain, or an interpolation between two or more atmospheric soundings. HPAC is dependent on the resolution, fidelity, and accuracy of the imported flow field. The HPAC methodology estimates the uncertainty in both the flow-field characterization and in the transport and dispersion operations done in those fields. Because HPAC is most generally used for operational planning based on a weather forecast, the quality of that forecast is the major limitation and the driver of uncertainties in modeling atmospheric transport and dispersion.

Personnel, Health, and Environmental Effects Codes

In health effects codes, fatalities are calculated on the basis of the number of people surviving at 60 days after a nuclear event. Injuries are calculated on the basis of incidence of illness from episodes of nausea and vomiting or unusual fatigability and weakness. Casualties are the sum of fatalities and injuries.[10] HPAC uses the Probability of Damage Calculator (PDCALC) to calculate the probability of fatalities and casualties, based on a population database, from prompt effects of a nuclear explosion. HPAC calculations of fallout employ the Radiation Induced Performance Distribution (RIPD), which provides human-effects estimations based on dose, dose rate, shielding, and time. Fatalities are calculated on the basis of a lognormal dose-response curve with a median lethal dose of 410 rads. Injury is calculated on the basis of early effects of exposure to radiation at a median effective dose of 210 rads (fatigability) and 240 rads (weakness). External dose is calculated based on the use of the assumption that exposure rate decreases with time $t^{-1.3}$. It estimates the response of military personnel in terms of physiological repair and/or recovery based on detailed models of relevant biological processes.

Transport and dispersion models estimate casualties by drawing dosage isopleths on the calculated or predicted curves of the downwind plume. Each model uses different methods to calculate the concentration in the plume, but essentially all calculate dosage by integrating the concentration at a point over time. Only HPAC includes a subroutine that calculates casualties from a nuclear event. HPAC calculates or predicts chemical and biological agent dosages and isopleths based either on program defaults or on user input.

HPAC uses the SCIPUFF model to predict concentrations of agents or radioactive particles and fluctuations in concentration. These predictions of airborne concentration are integrated over time to obtain an integrated dose of surface fallout, which is used to estimate lethality in human populations. Deposited mass on the surface may also be used to estimate percutaneous effects of exposure to chemical agents.

This probabilistic prediction of concentration or dose level can be integrated over the surface area to give an estimate of population exposure when the data on dose are coupled with data on population density. The result is an estimate of the number of people exposed to radiation or agent at a given dose or concentration. Conversely, an estimate of the area covered by radiation or agent at a particular concentration is also available, depending on the output that an operator selects. Estimation of a hazardous area is based on a calculation of dispersion that includes the uncertainties in the data input for wind conditions. Prediction of a hazardous area is based on an estimate that indicates the probability of exceeding a threshold value for concentration of radiation or agent.

SCIPUFF not only provides a mean value concentration, but also predicts a quantitative value for the random fluctuation in that value owing to turbulent diffusion. The results are a quantitative characterization of the validity of the prediction.

What Are the Major Limitations or Uncertainties?

HPAC's estimates of effects are for those deemed to be militarily significant. Hence, low-dose effects on humans, effects on the environment (crop production, fishing, and so on), and economic impacts are not addressed. Combined effects (e.g., increased susceptibility to disease after exposure to radiation) are not considered. Within the methodology, human-effects errors in HPAC estimates of casualties should be within a factor of two for military-age (17- to 24-year-old) male personnel who have no medical treatment or additional injuries. Effects on the full demographic range, from infants to the elderly and infirm, may vary significantly.

Databases

Several types of databases are utilized by these codes: population, agent fate, weather, and terrain.

Population Data

HPAC uses LandScan 2002, supplied by the Oak Ridge National Laboratory. Based on 1 kilometer by 1 kilometer worldwide grid cells, it is the only known high-resolution worldwide population database. The methodology for building LandScan population cells involves determining the probability that someone lives in a cell and the number of people in the cell. The probability of a cell's being inhabited is based on factors such as proximity to roads and water, land cover, elevation/slope, nighttime lights, and data from satellite imagery. Estimates of the number of people are based on the best census data from the U.S. Bureau of the Census and the International Programs Center. Data on land use and urban canopy are used for computations of ground-level atmospheric transport. Building inhabitant probability and building-type structural protection factors are derived to estimate probabilities of structural protection for each cell.

What are the major limitations or uncertainties? The LandScan population database does not account for events such as population movements (e.g., commuting), evacuation, or special events (e.g., the Olympics). The number of people in each cell can range over an order of magnitude when such factors are taken into account. Only detailed intelligence at a specific time can reduce this uncertainty. As with algorithms for predicting the effects of explosions, the population data contain no demographics.

Agent Fate

After they are released into the air, biological agents degrade as a function of the levels of radiant solar energy, relative humidity, temperature, ozone, and hydroxyl radicals. Biological agents are particularly susceptible to inactivation by the ultraviolet energy in sunlight; even anthrax spores degrade rapidly in bright daylight. Although some chemical agents are environmentally sensitive, neither GB (Sarin) nor VX is degraded significantly by sunlight, or by hydrolysis suspension in the atmosphere. For environmentally sensitive agents, the duration of exposure to sunlight and to photochemically produced compounds and radicals is an important factor in their loss of viability or toxicity. Biological agents released by explosive destruction of the sites where they are stored will be subject to inactivation by the high temperatures generated in the explosion, and chemical agents will be degraded to the extent that they are affected by very high temperatures. Those agents that do not readily aerosolize will contaminate the immediate surrounding area, where they will ultimately be degraded by hydrolysis or another chemical reaction.

What are the major limitations or uncertainties? Within the methodology for HPAC, agent decay is a minor source of uncertainty within the first 180 days after a nuclear burst or accident, or release of a biological or chemical agent. Beyond that point, long-lived radionuclides dominate. Since sunlight has an ultraviolet (UV) spectrum that is very effective in killing spores, HPAC uses a decay constant of 6 hours for spores.

Weather

HPAC incorporates many sources of weather data. With this variety comes the burden of determining the best possible data for a particular HPAC project. This selection depends on a number of factors, including the type of project (nuclear, radiological, chemical, or biological) and the timeline of the event (scenario planning, prestrike decision making, or poststrike analysis).

Different hazard scenarios require different types of meteorological data, which can come from observations or from forecast models. The meteorological data entered into HPAC may be categorized according to four types: (1) a single observation (fixed wind), (2) climatology (historical weather data), (3) surface and upper-air profile observations, and (4) weather model predictions in either gridded or profile format.

For a limited chemical release, a single, representative wind observation probably is sufficient to make a reasonable calculation of transport. Data for a fixed wind can be entered directly using a simple interface. Fixed wind is the simplest weather data type. HPAC assumes that a fixed wind is horizontally uniform. The fixed-wind option uses a theoretical vertical profile unless terrain is added. While a fixed wind is not the most accurate way of specifying weather conditions, the fixed-wind option does provide a quick calculation. This method can produce data to supplement climatology data when there are no available climatic data or when the goal is to display incident plumes driven by winds from varying directions.

Historical weather data created by the Air Force Combat Climatology Center from multiyear records of weather data are included on the HPAC CD-ROM. These historical weather data include the effects of both time of day (diurnal) and seasonal variations on the weather. The data sets provide quantitative meteorological input for long-range planning and incidents for which no other weather information is available.

HPAC includes both surface and upper-air historical weather data. Surface weather data are most useful for chemical incidents with durations of 4 hours or less. Upper-air weather data are more representative for larger domains and for longer-duration hazards such as those produced by the use of nuclear weapons or the release of biological agents, or for higher-altitude releases of materials (releases above 500 meters). Whenever historical weather data are selected as the weather input, HPAC defaults to using the upper-air historical weather data.

What are the major limitations or uncertainties? Several papers have recently appeared in the literature on the use of the ensemble method with transport and turbulent diffusion models. Many groups are currently carrying out sensitivity studies to determine how the ensemble approach can best be implemented to improve their forecasts. In this approach, *N* meteorological models with several alternate inputs or initialization options are run on the same scenario. At the moment, *N* typically ranges from 5 to 20. It is found that the average of *N* forecasts is more accurate than any single-model forecast. The forecasters try to make sure that the spread or uncertainty range for *N* forecasts is approximately the same as the spread in the climatological observations. This is an empirical approach, since there is no a priori reason that the spread of the ensemble should equal the observed spread. Use of the ensemble method is growing in the meteorological and air-quality modeling area, but it has not reached maturity. Research is needed to put the ensemble method on a better statistical and scientific basis, especially concerning the spread in or uncertainty of the results.

Terrain

HPAC uses digital topographic elevation data (DTED) Level 0 terrain-elevation data developed by the National Geospatial-Intelligence Agency (NGA). DTED Level 0 elevation data have post spacing of 30 arc seconds (nominally 1 kilometer). These data allow construction of a uniform matrix of terrain-elevation values that provide basic quantitative data for systems and applications that require terrain-elevation, slope, and/or surface roughness information. DTED Level 0 is derived from the higher-resolution NGA DTED Level 1 data. Level 1 data have post spacing of 3 arc seconds (nominally 100 meters). Higher-resolution data are available. Balancing calculational accuracy with run-time requirements drove the choice of DTED Level 0 as the default data.

Run-time calculations of atmospheric transport and dispersion (ATD) require longer run times when higher-resolution terrain data are used. Some operational users require answers quickly and can decrease run-time demands by reducing the number of grid posts in the calculational domain, or by changing the domain, or both. For a particular domain size, several choices of resolution are available, ranging from the native resolution to resolution as little as 900 points over the calculational domain. Another option is to not use any terrain-elevation data (sometimes called a flat-earth approach) in the calculation.

HPAC defaults to different domain and corresponding terrain resolutions depending on the simulation scenario. A long-ranging nuclear fallout hazard scenario will cause HPAC to default to use of a large domain, whereas a chemical event that is driven by local winds has a smaller domain. The user can override all of the defaults but must be careful to balance mission requirements (e.g., calculational run time) with output accuracy.

As use of ATD tools has begun to address hazards in the urban environment, the concept of terrain has been extended to include urban terrain. Data on urban terrain represent the building location and height information. In contrast to the worldwide DTED data, digitized data on urban locales, compatible for input into urban transport and dispersion models, are available for only a handful of cities. There is currently no federal point of contact responsible for developing these data worldwide—that is, the NGA equivalent for terrain elevation. It is important to add, although it is beyond the scope of this discussion, that some urban models do not use building height data. These models extend the traditional (nonurban) ATD models by altering wind-flow assumptions and characterizing the urban environment as a surface roughness.

Terrain affects wind flow. In the presence of complex terrain, wind tends to flow around hills and to change speed. The height of the plume above the local surface changes as it passes along a hill or other elevation features. Both of these effects alter the local wind direction as well as the downwind transport and turbulent diffusion processes. The ATD models account for terrain. Numerical weather models (input to ATD models) factor in terrain as part of their calculational boundary conditions, but the calculations are performed on a coarse scale. These models typically predict wind information at resolutions of 10s to 100s of kilometers and can be run at several different resolutions, but long run times and associated resources are required. Even real-time observations of wind are spatially separated by significant terrain and are separated in time.

HPAC uses the Stationary Wind Fit and Turbulence (SWIFT) mass-consistent wind model to extrapolate the coarser-resolution wind data (from forecasts or observations) to a finer resolution, in which terrain features affect wind flow. When terrain is not a factor in the local wind flow or because run times demand a quick-look assessment, then another model, the Mass Consistent SCIPUFF (MC-SCIPUFF), is used by HPAC.

What are the major limitations or uncertainties? The data on terrain and elevation are available at several levels of spatial resolution, as discussed above. The DTED terrain elevation data are quality-controlled, and the precision of the data is usually not an issue. However, spatial resolution can be a large factor in determining the accuracy of the ATD calculation and downwind hazard.

The resolution of the terrain data used should be matched to the simulation scenario. Hazard predictions of continental-scale radiation fallout from large nuclear weapon incidents do not require highly resolved data on surface winds. Terrain-elevation data can be coarse, and in some locations a flat-earth approximation is sufficient. Although ATD models can be run with high-resolution terrain data, at some point the user is getting a false precision at the expense of very long run times. A longer-running ATD code does not mean a more accurate ATD code. When hazards must be characterized near the source as in the example of chemical or biological releases, the terrain is important.

The HPAC model can use very high spatial resolution terrain-elevation data. The accuracy of the prediction is a trade-off between mission requirements supporting an actionable decision and operational constraints associated with run-time demands.

NOTES

1. See the subsection "Databases" in Attachment 5.1 for a description.

2. Samuel Glasstone and Philip J. Dolan (eds.). 1977. *The Effects of Nuclear Weapons*, U.S. Government Printing Office, Washington, D.C., §2.18–2.21, §2.23–2.31, §2.91–2.95, §2.99–2.100.

3. Todd Hann, Defense Threat Reduction Agency, August 13, 2004, personal communication.

4. Todd Hann, Defense Threat Reduction Agency, August 13, 2004, personal communication.

5. Todd Hann, Defense Threat Reduction Agency, August 13, 2004, personal communication.

6. Matthew Meselson, Jeanne Guillemin, Martin Hugh-Jones, Alexander Langmuir, Llona Popova, Alexis Shelokov, and Olga Yampolskaya. 1994. "The Sverdlovsk Anthrax Outbreak of 1979," *Science*, Vol. 266, pp. 1202-1208.

7. National Research Council. 2003. *Tracking and Predicting Atmospheric Dispersion of Hazardous Material Releases: Implications for Homeland Security*, The National Academies Press, Washington, D.C.

8. C. du P. Donaldson. 1973. *Atmospheric Turbulence and the Dispersal of Atmospheric Pollutants,* AMS Workshop on Micrometeorology, D.A. Haugen (ed.), Science Press, Boston, Mass., pp. 313-390.

9. W.S. Lewellen. 1977. "Use of Invariant Modeling," *Handbook of Turbulence*, W. Frost and T.H. Moulden (eds.), Plenum Press, pp. 237-280.

10. A soldier is considered injured if s/he is unable to perform.

6

Human and Environmental Effects

EFFECTS ON HUMANS

The health effects of nuclear explosions are due primarily to air blast, thermal radiation, initial nuclear radiation, and residual nuclear radiation or fallout.

- *Blast.* Nuclear explosions produce air-blast effects similar to those produced by conventional explosives. The shock wave can directly injure humans by rupturing eardrums or lungs or by hurling people at high speed, but most casualties occur because of collapsing structures and flying debris.
- *Thermal radiation.* Unlike conventional explosions, a single nuclear explosion can generate an intense pulse of thermal radiation that can start fires and burn skin over large areas. In some cases, the fires ignited by the explosion can coalesce into a firestorm, preventing the escape of survivors. Though difficult to predict accurately, it is expected that thermal effects from a nuclear explosion would be the cause of significant casualties.
- *Initial radiation.* Nuclear detonations release large amounts of neutron and gamma radiation. Relative to other effects, initial radiation is an important cause of casualties only for low-yield explosions (less than 10 kilotons).
- *Fallout.* When a nuclear detonation occurs close to the ground surface, soil mixes with the highly radioactive fission products from the weapon. The debris is carried by the wind and falls back to Earth over a period of minutes to hours.

The first three of these effects are "prompt" effects, because the harm is inflicted immediately after the detonation. By contrast, the radiation dose from fallout is delivered over an extended period, as described in Chapter 5. Most of the dose from fallout is due to external exposure to gamma radiation from radionuclides deposited on the ground, and this is the only exposure pathway considered by the computer models that the Defense Threat Reduction Agency (DTRA) and Lawrence Livermore National Laboratory (LLNL) used to estimate health effects for this study. Below is a discussion of the possible contribution of other exposure pathways, such as inhalation of contaminated air and consumption of contaminated water and food, to the total radiation dose received by humans.

Radiation has both acute and latent health effects. Acute effects include radiation sickness or death resulting from high doses of radiation (greater than 1 sievert [Sv], or 100 rems) delivered over a few days. The principal latent effect is cancer. Estimates of latent cancer fatalities are based largely on results of the long-term follow-up of the survivors of the atomic bombings in Japan. The results of these studies have been interpreted by the International Commission on Radiological Protection (ICRP)[1] in terms of a lifetime risk coefficient of 0.05 per sievert (5×10^{-4} per rem), with no threshold.[2] For the present study, acute radiation effects were estimated by both DTRA and LLNL; latent cancer deaths were estimated only by LLNL.

The computer models used by DTRA and LLNL were developed primarily to estimate effects on military personnel rather than for civilian populations. Thus, there is no consideration of the presumed greater sensitivity to radiation of the very young and the elderly. Also, there is no consideration of the sensitivity of the fetus. From the experience in Japan, it is known that substantial effects on the fetus can occur, and these effects depend on the age (stage of organogenesis) of the fetus.[3] One such effect is mental retardation. The transfer of radio nuclides to the fetus resulting from their intake by the mother is another pathway of concern. Radiation dose coefficients for this pathway have been published by the ICRP.[4]

Another long-term health effect that is not considered here is the induction of eye cataracts. This effect has been noted in the Japanese studies and also in a study of the Chernobyl cleanup workers.[5]

Compared to the fatalities from prompt, acute fallout and latent cancer fatalities, the absolute number of effects on the fetus is small and is captured within the bounds of the uncertainty. The number of eye cataracts, based on the experience of the Chernobyl workers, is not small. The occurrence of eye cataracts in the now aging Japanese population is several tens of percent among those more heavily exposed.

Finally, there has been a recently confirmed finding that the Japanese survivors are experiencing a statistically significant increase in the occurrence of a number of noncancer diseases,[6] including hypertension, myocardial infarction, thyroid disease, cataracts, chronic liver disease and cirrhosis, and, in females, uterine myoma. There has been a negative response in the occurrence of glaucoma. A nominal risk coefficient for the seven categories of disease is about 0.9 Sv^{-1} (0.009 rem^{-1}). The largest fraction of the risk is due to thyroid disease.

Thermal Radiation from Underground Bursts

Thermal radiation may make fire a collateral effect of the use of surface burst, airburst, or shallow-penetrating nuclear weapons. The potential for fire damage depends on the nature of the burst and the surroundings. If there is a fireball, fires will be a direct result of the absorption of thermal radiation. Fires can also result as an indirect effect of the destruction caused by a blast wave, which can, for example, upset stoves and furnaces, rupture gas lines, and so on. A shallow-penetrating nuclear weapon of, say, 100 to 300 kilotons at a 3 to 5 meter depth of burst will generate a substantial fireball that will not fade as fast as the air blast.

Detonation of a nuclear weapon in a forested area virtually guarantees fire damage at ranges greater than the range of air-blast damage. If the burst is in a city environment where buildings are closely spaced, say less than 10 to 15 meters, fires will spread from burning buildings to adjacent ones. In Germany and Japan in World War II, safe separation distance ranged from about 30 to 50 feet (for a 50 percent probability of spread), but for modern urban areas this distance could be larger. This type of damage is less likely to occur in suburban areas where buildings are more widely separated.

Once started, fire spread continues until the fire runs out of fuel or until the distance to the next source of fuel is too great. Thus, fire caused directly by thermal ignitions, fire caused indirectly by disruptive blast waves, and spread of fire are all potential, but uncertain, effects.

Illustrative Example: Washington, D.C.

The area over which casualties would occur as a result of the various weapon effects outlined above depends primarily on the explosive yield of the weapon and the height or depth of the burst. The areas affected by initial nuclear radiation and fallout also depend on the design of the weapon (in particular, the fraction of the yield that is derived from fission reactions), and, in the case of fallout, on weather conditions during and after the explosion (notably wind speed and direction, atmospheric stability, precipitation, and so on), terrain, and geology in the area of the explosion. The following calculations assume that the entire population is static and in the open.

As an illustrative example,[7] Figure 6.1 shows the area over which an individual in the open would face a 10, 50, and 90 percent chance of death or serious injury[8] from the prompt effects of a 10-kiloton earth-penetrator weapon (EPW) detonated at a depth of 3 meters and from the prompt effects of a 250 kiloton surface burst. (As discussed in Chapter 5, both of these weapons would produce a ground shock of about 1 kilobar at a depth of 70 meters.) These contours, which were produced by the DTRA using the Hazard Prediction and Assessment Capability (HPAC) code, are shown on a map of Washington, D.C., for scale. Figure 6.2 is similar, but also includes the probability of death or serious injury from acute exposure to external gamma radiation from fallout, for illustrative weather conditions, assuming hypothetically that 50 percent of the weapon yield is derived from fission and that a static population is in the open.

Figure 6.3 compares the numbers of casualties (deaths and serious injuries) due to prompt and acute effects of fallout from the use of both weapons. Under these conditions and assumptions, the 10 kiloton EPW is estimated to result in about 100,000 casualties, compared with 800,000 casualties for the

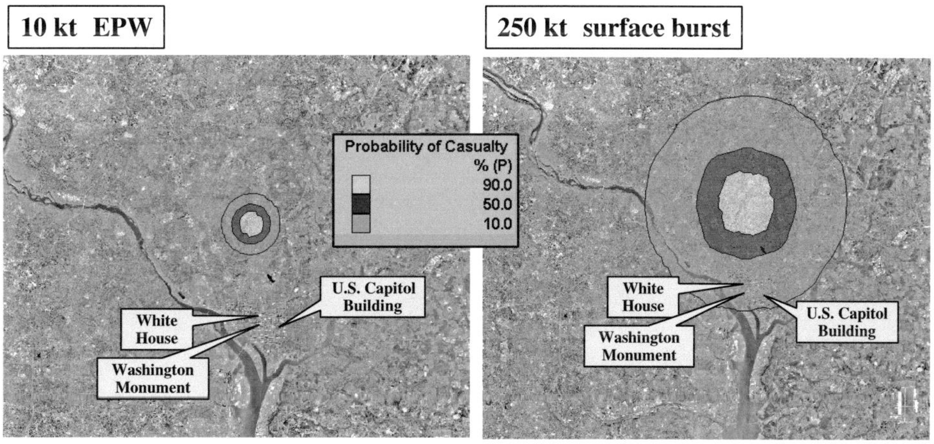

FIGURE 6.1 Illustrative example: The area over which an individual in the open would face a 10, 50, and 90 percent chance of death or serious injury from the prompt effects of a 10 kiloton earth-penetrator weapon (EPW; left) and a 250 kiloton surface burst (right) detonated at 7:00 p.m. on July 14, 2004, in Washington, D.C. SOURCE: Estimates prepared for the committee by the Defense Threat Reduction Agency.

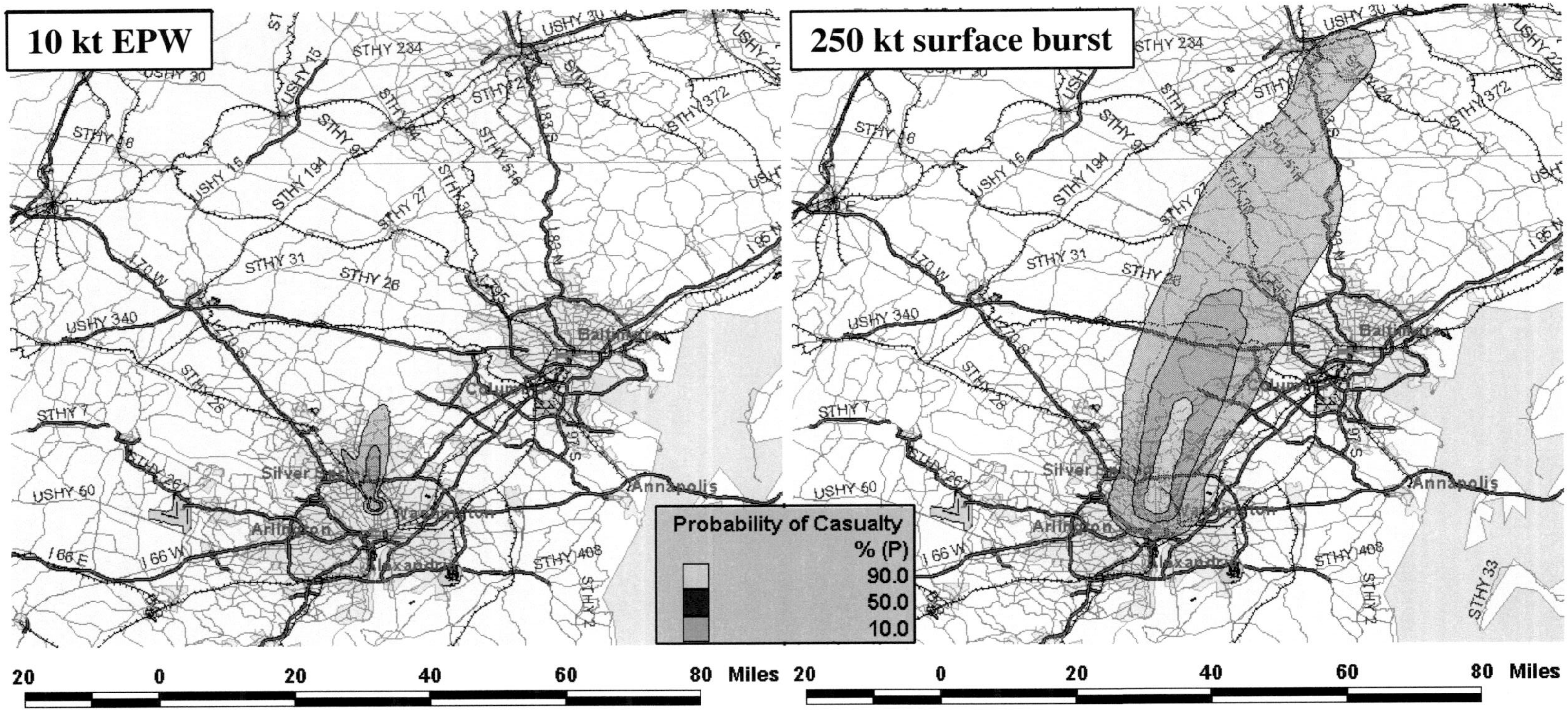

FIGURE 6.2 Illustrative example: The area over which an individual in the open would face a 10, 50, and 90 percent chance of death or serious injury from the prompt effects of fallout from a 10 kiloton earth-penetrator weapon (EPW; left) and a 250 kiloton surface burst (right) detonated at 7:00 p.m. on July 14, 2004, in Washington, D.C. SOURCE: Estimates prepared for the committee by the Defense Threat Reduction Agency.

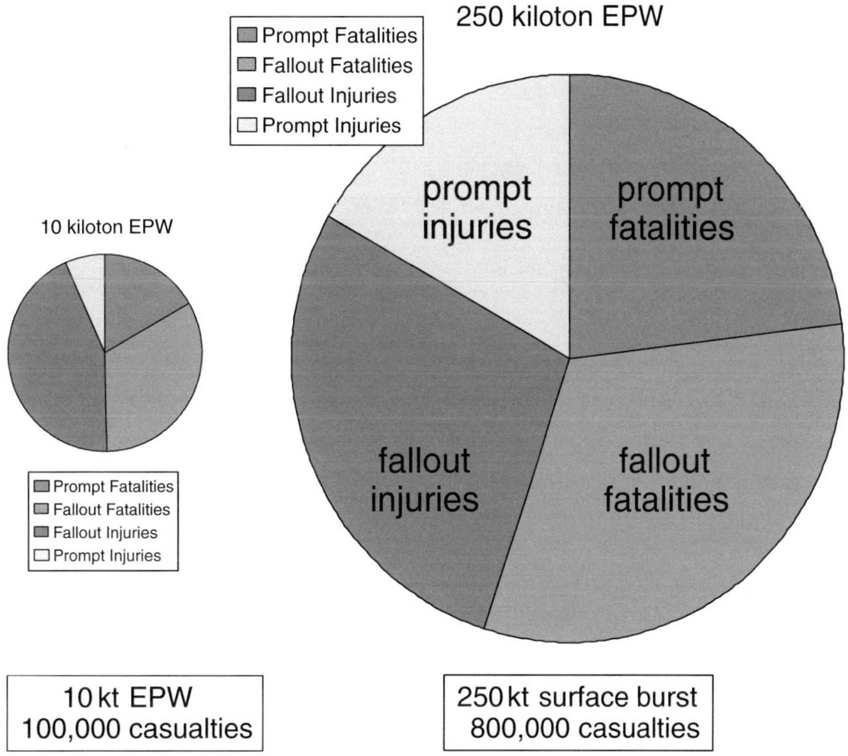

FIGURE 6.3 Illustrative example: Comparison of the number of casualties (deaths and serious injuries) from prompt and acute effects of fallout from a 10 kiloton earth-penetrator weapon (EPW) and a 250 kiloton surface burst detonated at 7:00 p.m. on July 14, 2004, in Washington, D.C. SOURCE: Estimates prepared for the committee by the Defense Threat Reduction Agency.

250 kiloton surface burst. Thus, in this example the use of an EPW would reduce casualties by about a factor of eight compared with a surface burst with equal destructive capacity against a buried target. Fallout is responsible for about 75 percent of the casualties from the 10 kiloton explosion compared with about 60 percent of the casualties from the 250 kiloton explosion.

The hazard to people entering the area after the explosion in these scenarios would be due largely to external gamma radiation from fallout. This hazard decreases rapidly with time: the dose rate after 1 week is 10 times less than the dose rate 1 day after the explosion, and after 2 months it is reduced by an additional factor of 10. Figures 6.4 and 6.5 illustrate this decay for the cases described above (the 10 kiloton EPW and the 250 kiloton surface burst), respectively, showing the areas exceeding a dose rate of 0.01, 0.1, 1, and 10 millisieverts per hour (1, 10, 100, and 1,000 millirems per hour) at 1 day, 1 week, 1 month, and 6 months after the explosion.

To put the dose rates referred to above in perspective, a person who remained indefinitely in an area where the dose rate was 1 millirem per hour at the time of that person's entry into the area would receive a total dose of less than 50 millisieverts (5 rems), which is the annual dose limit for U.S. nuclear workers.[9] Thus, military personnel could enter the unshaded areas shown in Figures 6.4 and 6.5 at the times indicated with minimal risk. Depending on the risk that is judged acceptable by commanders,

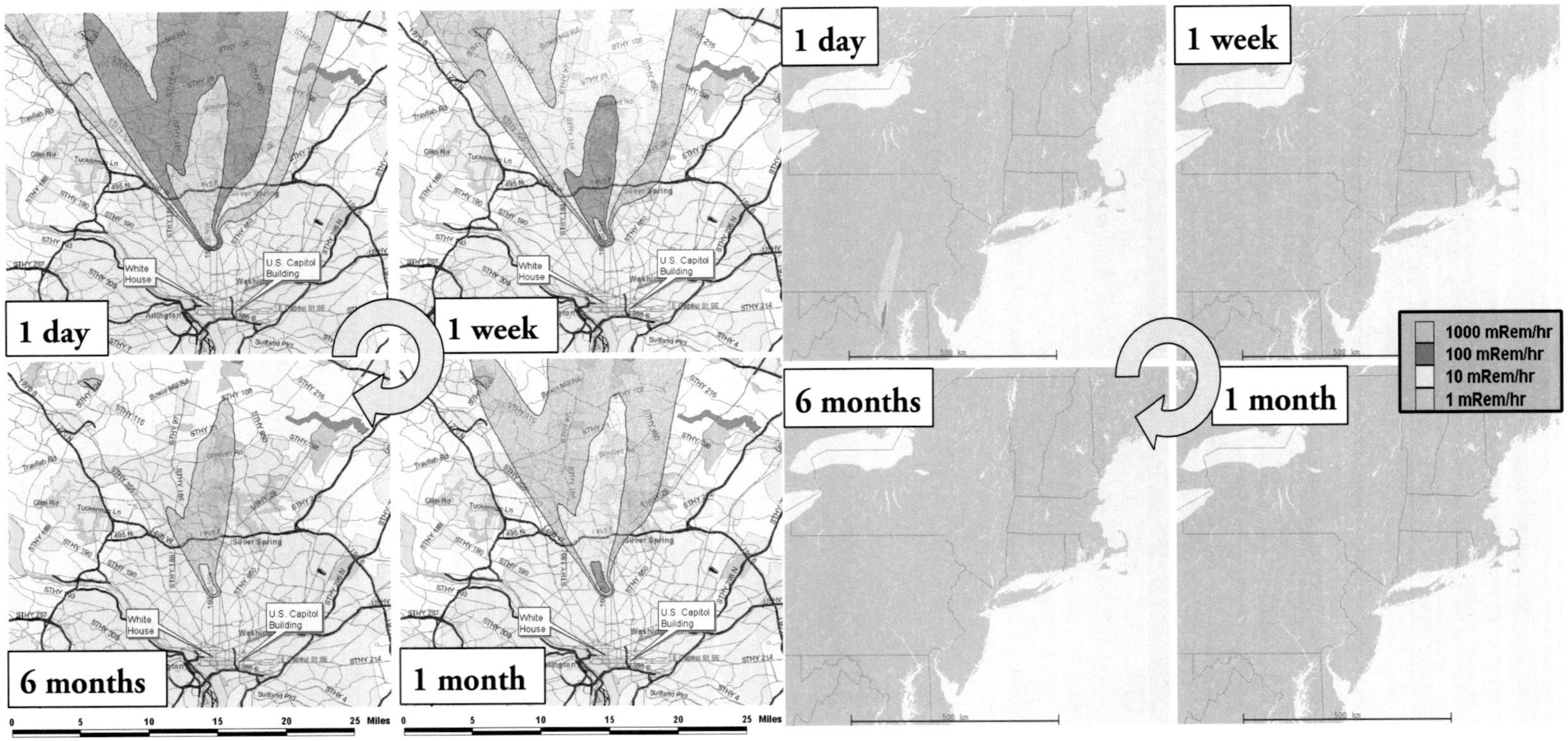

FIGURE 6.4 Illustrative example: Areas within which the dose rate from external gamma radiation exceeds 1, 10, 100, and 1,000 millirems per hour at 1 day, 1 week, 1 month, and 6 months after the detonation of a 10 kiloton earth-penetrator weapon at 7:00 p.m. on July 14, 2004, in Washington, D.C. SOURCE: Estimates prepared for the committee by the Defense Threat Reduction Agency.

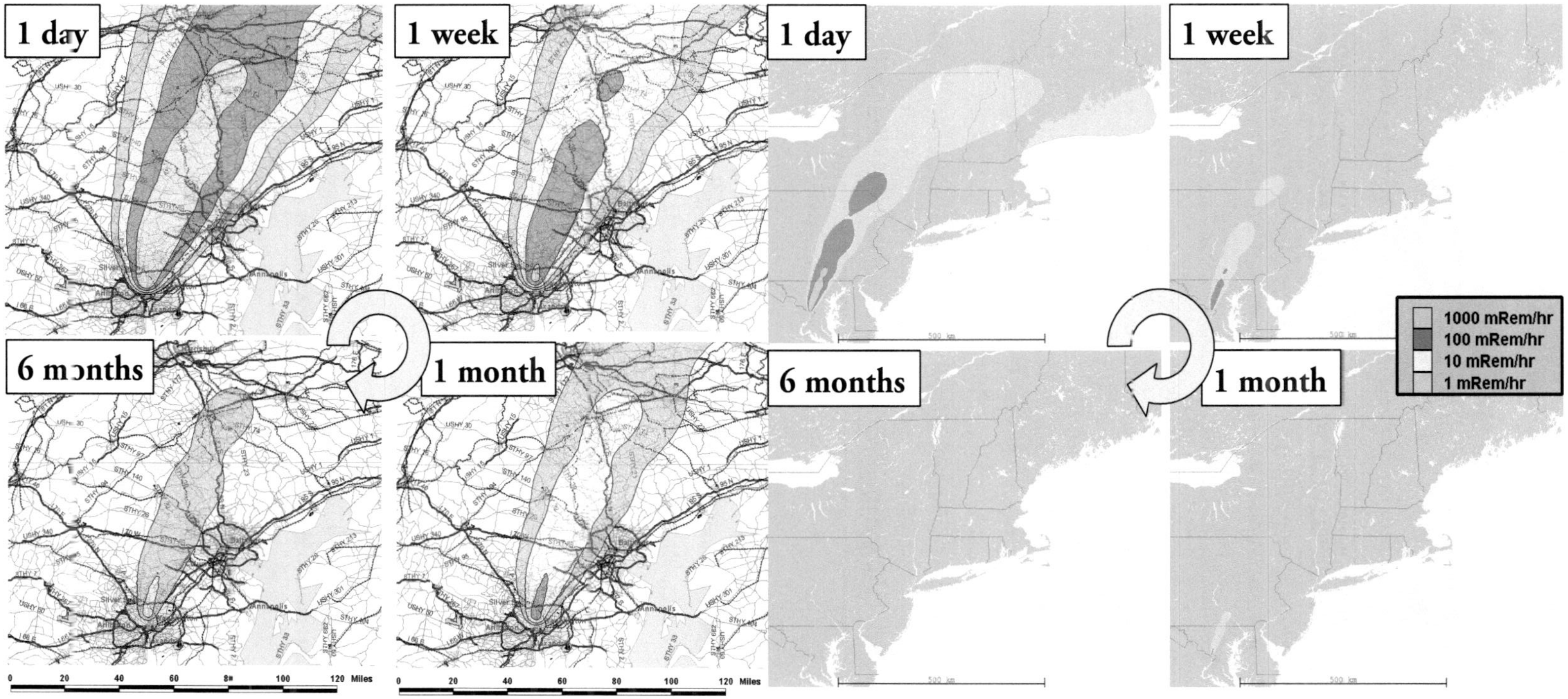

FIGURE 6.5 Illustrative example: Areas within which the dose rate from external gamma radiation exceeds 1, 10, 100, and 1,000 millirems per hour at 1 day, 1 week, 1 month, and 6 months after the detonation of a 250 kiloton surface burst at 7:00 p.m. on July 14, 2004, in Washington, D.C. SOURCE: Estimates prepared for the committee by the Defense Threat Reduction Agency.

soldiers could enter shaded areas for various periods of time. For example, a soldier entering the 10 millisieverts per hour (1,000 millirems per hour) contour 1 day after the explosion would accumulate a total dose of about 0.25 sievert (25 rems) over the next 2 days and 0.50 sievert (50 rems) over the next 2 weeks. U.S. regulatory guidelines allow doses of up to 0.25 sievert (25 rems) in lifesaving emergency situations, and the U.S. National Council on Radiation Protection and Measurements (NCRP) recommends that doses up to 0.50 sievert (50 rems) be allowed in such situations provided that individuals are aware of the risks.[10] Two reports from the Institute of Medicine address the U.S. Army guidance for situations in which troops might receive as much as 0.70 sievert (70 rems).[11] Doses of 0.25 to 0.70 sievert (25 to 70 rems) are unlikely to cause serious acute effects, but they may ultimately cause death due to cancer in 1 to 3 percent of those exposed (in addition to the roughly 20 percent lifetime risk of dying of cancer from other causes).

Other Targets and Weapons Yields[12]

The estimates shown in Figures 6.1 through 6.5 apply only to a particular set of assumptions about target location, weather, and weapons used to attack the target. The number of civilian casualties that would result from an attack depends on many variables, including the following: the distribution of the population around the point of detonation and the degree of sheltering that they have against blast, thermal, and radiation effects; weapon yield and design; height or depth of burst; and weather conditions during and after the explosion. As shown below, the estimated number of casualties ranges over four orders of magnitude—from hundreds to over a million—depending on the combination of assumptions used.

To explore in a parametric way the range of possibilities, the committee selected three notional targets:

- *Target A:* an underground command-and-control facility in a densely populated area 3 kilometers from the center of a city with a population of about 3 million;
- *Target B:* an underground chemical warfare facility 60 kilometers from the nearest city and 13 kilometers from a small town; and
- *Target C:* a large, underground nuclear weapons storage facility 20 kilometers from a small town.

In each case, the committee asked DTRA to estimate the mean number of casualties (deaths and serious injuries from prompt effects, and acute effects of fallout from external gamma radiation) resulting from attacks with earth-penetrating weapons with yields ranging from 1 kiloton to 1 megaton, for populations completely in the open and completely indoors. The means are averages over annual wind patterns, but they ignore precipitation. DTRA also estimated the mean number of casualties resulting from surface bursts with yields from 25 kilotons to 7.5 megatons. For selected cases, the committee asked the Lawrence Livermore National Laboratory to estimate the number of deaths from prompt effects and fallout, and to quantify the variability in acute and latent deaths from fallout owing to wind patterns.

For Figures 6.6 and 6.7 the calculations assume that the entire population is static and in the open. Figure 6.6 shows the estimated mean number of casualties resulting from attacks on Targets A, B, and C with surface-burst weapons and earth-penetrator weapons of a range of yields from 1 kt to 10 kt, with the EPW detonated at a depth of 3 meters, assuming a static population in the open. Note that for a given yield there is little or no difference between the effects of surface bursts and the EPWs.[13] The curves for Target A are relatively flat (a factor-of-10 increase in yield produces a factor-of-2 increase in casualties) because the population is clustered around the target. The curves for Targets B and C are steeper (a

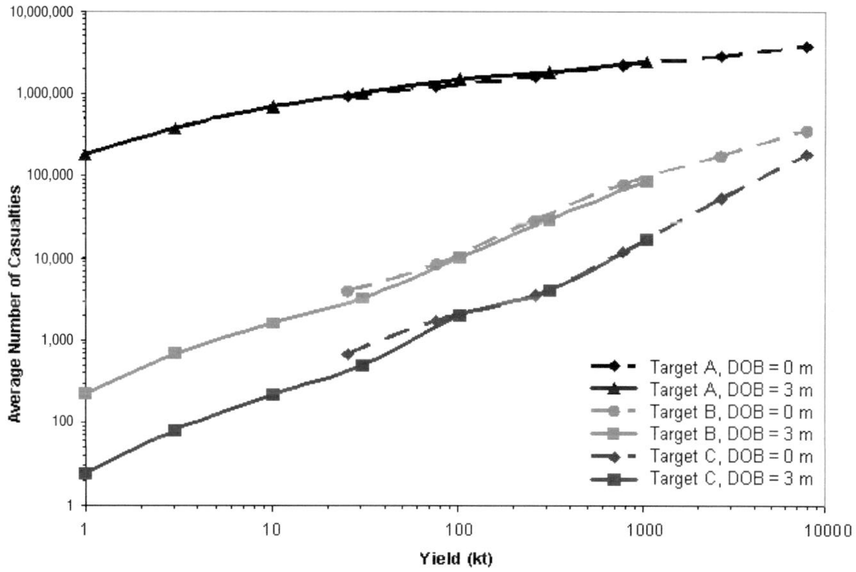

FIGURE 6.6 Estimated mean number of casualties (deaths and serious injuries) from attacks on notional targets A, B, and C using earth-penetrator weapons at 3 meters' depth of burst and surface bursts, assuming a static population in the open. SOURCE: Estimates prepared for the committee by the Defense Threat Reduction Agency.

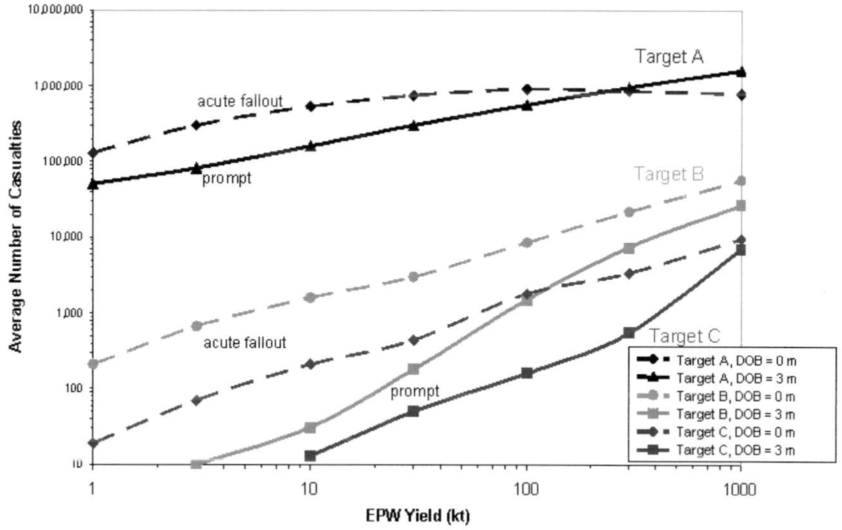

FIGURE 6.7 Estimated mean number of casualties from prompt effects and acute radiation sickness and death from fallout resulting from attacks on notional targets A, B, and C using earth-penetrator weapons (EPWs) at 3 meters' depth of burst, assuming the entire population is in the open. SOURCE: Estimates prepared for the committee by the Defense Threat Reduction Agency.

factor-of-10 increase in yield produces a factor-of-6 to -10 increase in casualties), largely because the effects from higher-yield explosions can reach more-distant population centers.

Figure 6.7 shows the contributions of prompt effects and acute radiation sickness and death from fallout to the casualty estimates for EPWs. (The number of casualties is similar for surface bursts of the same yield.) Note that for yields of less than 300 kilotons, fallout is responsible for more casualties than are prompt effects. This is particularly true for Targets B and C, for which fallout is the only effect of low-yield explosions that can reach population centers.

It is always useful to compare model predictions against relevant experience. Fortunately, the relevant experience is very limited. In the case of the 15 kiloton device detonated over Hiroshima, an estimated 68,000 persons died and 76,000 persons were injured out of a total population of 250,000. For the 21 kiloton device detonated over Nagasaki, it is estimated that 38,000 persons died and 21,000 persons were injured out of a total population of 170,000.[14] These estimates are in rough agreement with the estimated 200,000 prompt-effects casualties shown in Figure 6.7 for Target A, taking into account differences in the size of the vulnerable populations. (The Hiroshima and Nagasaki weapons were detonated at a fallout-free height of about 500 meters and therefore produced no local fallout.)

As mentioned, the results shown in Figures 6.1 through 6.7 assume that the entire population is static and in the open. Assuming that the entire population remains indoors and is thereby shielded from radiation reduces mean total casualties by a factor of up to 4 for Target A, and by a factor of 2 to 8 for Targets B and C. Not accounted for are post attack movement or evacuation of the population, but it is unlikely that individuals could, by fleeing the area of an attack, reduce their exposure to fallout significantly more than by remaining indoors. Indeed, some people might greatly increase their exposure to fallout if they were to move through highly contaminated areas, as might occur if a major road out of the city were directly under the path of the cloud. Thus, in a population that has received no warning of an attack, the actual effects of sheltering and evacuation are likely to lie between the two extremes for a population that is assumed to be entirely indoors and one that is assumed to be entirely outdoors.

The use of an EPW instead of a surface-burst weapon generally will result in fewer casualties, because the yield of the EPW can be 15 to 25 times smaller than the yield of a surface-burst weapon for a given level of damage against a hard and deeply buried target (HDBT). Figure 6.8 shows the ratio of the mean number of casualties estimated for a surface burst to the mean number estimated for an EPW with a yield 25 times smaller, for Targets A, B, and C. For Target A, casualties are reduced by a factor of 7 at low yields appropriate for target depths of less than 100 meters and by a factor of 2 at high yields and deeper targets. For Target B, casualties are reduced by a factor of 10 to 30, and for Target C, by a factor of 15 to 60, depending on the yield and assumptions about shielding. In general, the reduction factor is larger for targets in rural or remote areas.

The DTRA results presented above do not include latent cancer deaths from fallout. The committee asked LLNL to estimate the mean number of latent cancer deaths for Targets A and B, for yields from 10 to 300 kilotons.[15] In the case of Target A, the inclusion of latent cancer deaths increased the total estimated number of fatalities by less than 20 percent. In the case of Target B, however, the inclusion of cancer deaths doubled the total number of fatalities. Including cancer deaths has little effect on the ratios shown in Figure 6.8.

The results given in Figures 6.6 through 6.8 are averages over annual wind patterns. Casualties from fallout can be substantially higher or lower, depending on the particular wind conditions during and immediately following the attack. Figures 6.9(a) and (b) show the variation in the number of deaths due to acute and latent effects from fallout from a 300 kiloton EPW on Targets A and B, respectively, as a function of wind direction. For Target A, estimated fatalities from fallout vary by more than an order of magnitude depending on wind direction, ranging from 90,000 to 800,000 for acute effects and from

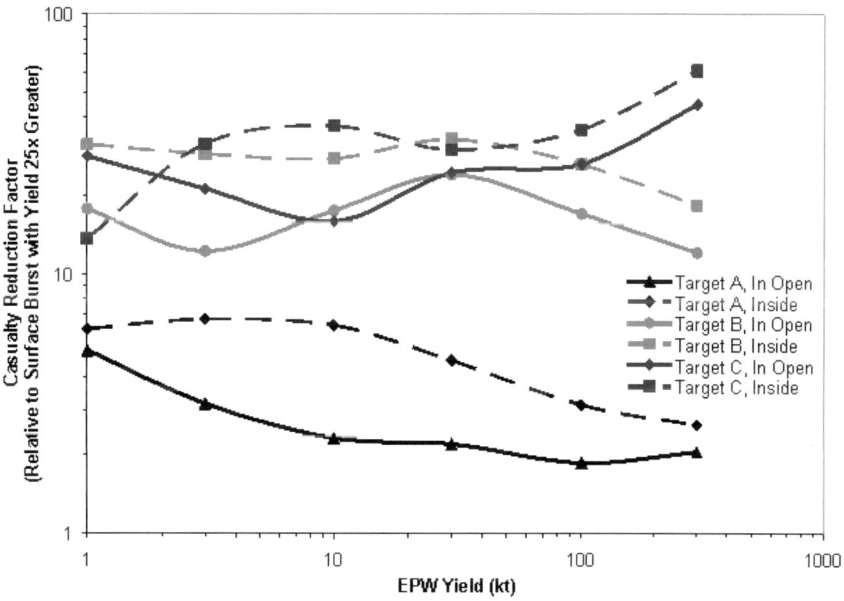

FIGURE 6.8 Ratio of the estimated mean number of casualties from a surface burst to the mean number from an earth-penetrator weapon (EPW) with a yield 25 times smaller, for notional targets A, B, and C, assuming a static population entirely in the open or entirely indoors. SOURCE: Estimates prepared for the committee by the Defense Threat Reduction Agency.

30,000 to 200,000 for latent effects; total fatalities, however, vary by less than a factor two, from 1 million to 2 million. For Target B estimated fatalities from fallout vary by more than two orders of magnitude depending on wind direction, from 3,000 to 1 million for acute fatalities, and ranging from 3,000 to 300,000 for latent fatalities; total fatalities vary by a factor of 50, from about 15,000 to 800,000. Similarly large variations in fatalities are also possible if the target is just outside a major city. For example, if the detonation is moved 30 kilometers northwest of Target A (hereafter referred to as Target A), total fatalities vary from 50,000 to nearly 2 million, depending on whether the wind blows away from or toward the city center. Note that these estimates do not include the effects of precipitation, which would wash out and concentrate fallout in particular areas (which may or may not be populated). The committee expects that including the effects of precipitation would make the weather-related variability in the estimated number of casualties significantly greater than is suggested by this analysis. Of course, as mentioned frequently, Figure 6.9(a) and (b) are model runs and therefore are subject to the sources of uncertainty described in this report and emphasized in Chapter 8.

Figures 6.10(a) and (b) use the information in Figures 6.9(a) and (b), together with the likelihood that the wind blows in each direction, to compute the probability of exceeding a given number of deaths due to acute and latent effects from fallout, as well as from all effects, for attacks with a 300 kiloton EPW on Targets A and B. In the case of Target A, for example, the 50 percent confidence interval for deaths due to acute effects of fallout (based solely on variability in wind direction) is 130,000 to 600,000; that is, there is a 75 percent chance of exceeding 130,000 deaths from acute effects of fallout, and a 25 percent chance of more than 600,000 deaths. The 50 percent confidence interval for total fatalities is considerably narrower: 1.1 million to 1.6 million. If the detonation is moved 30 kilometers northwest of Target A, the confidence intervals are much wider: 13,000 to 700,000 for deaths from acute

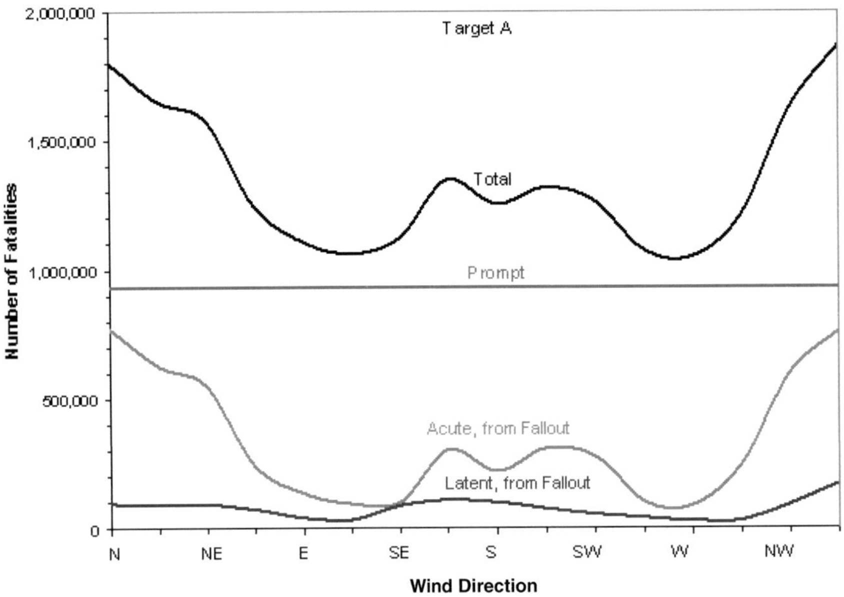

FIGURE 6.9(a) Variation in the estimated number of fatalities due to acute and latent effects from external gamma radiation from fallout from a 300 kiloton earth-penetrator weapon at 3 meters' depth of burst on notional target A as a function of wind direction, assuming that the population is in the open. SOURCE: Estimates prepared for the committee by the Lawrence Livermore National Laboratory.

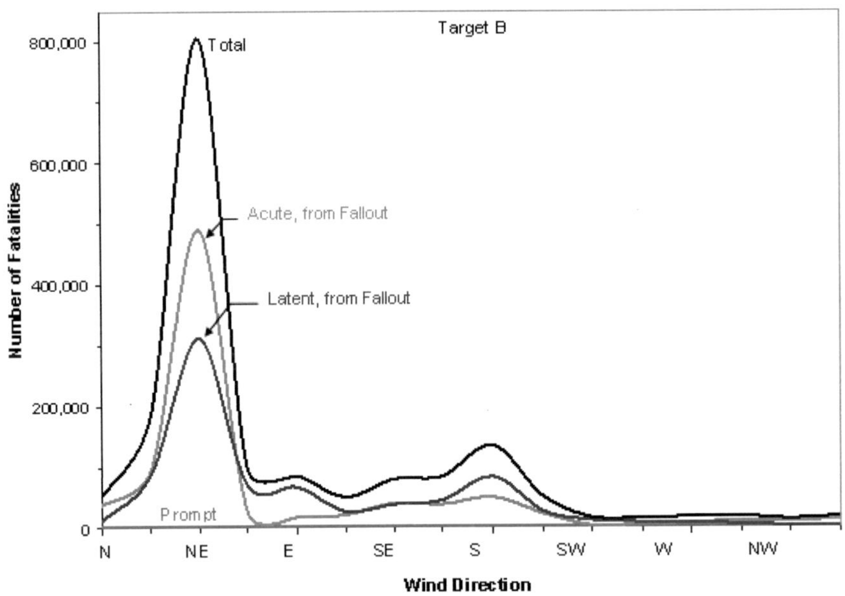

FIGURE 6.9(b) Variation in the estimated number of fatalities due to acute and latent effects from external exposure to gamma-radiation fallout from a 300 kiloton earth-penetrator weapon at 3 meters' depth of burst on notional target B as a function of wind direction, assuming that the population is in the open. SOURCE: Estimates prepared for the committee by the Lawrence Livermore National Laboratory.

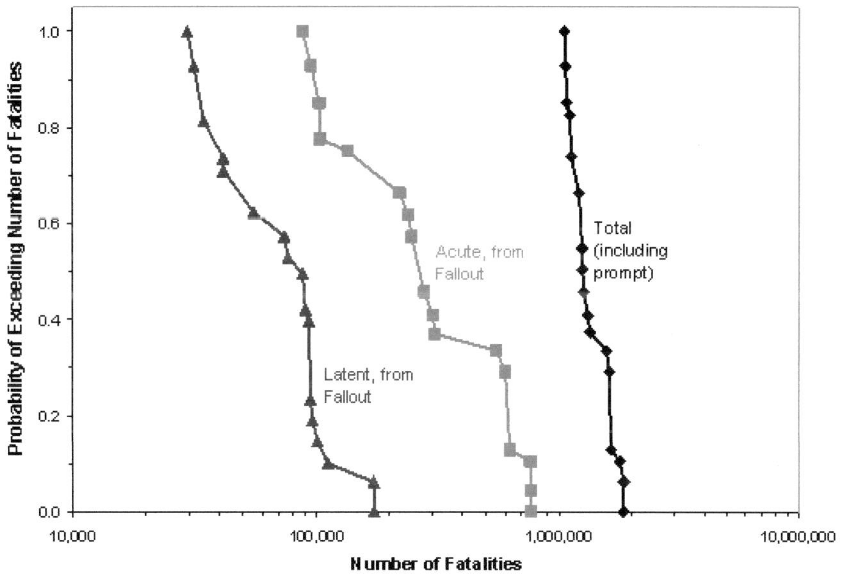

FIGURE 6.10(a) The probability of exceeding a given number of deaths due to acute and latent effects from external exposure to gamma-radiation fallout from a 300 kiloton earth-penetrator weapon at 3 meters' depth of burst on notional target A, assuming that the population is in the open. SOURCE: Estimates prepared for the committee by the Lawrence Livermore National Laboratory.

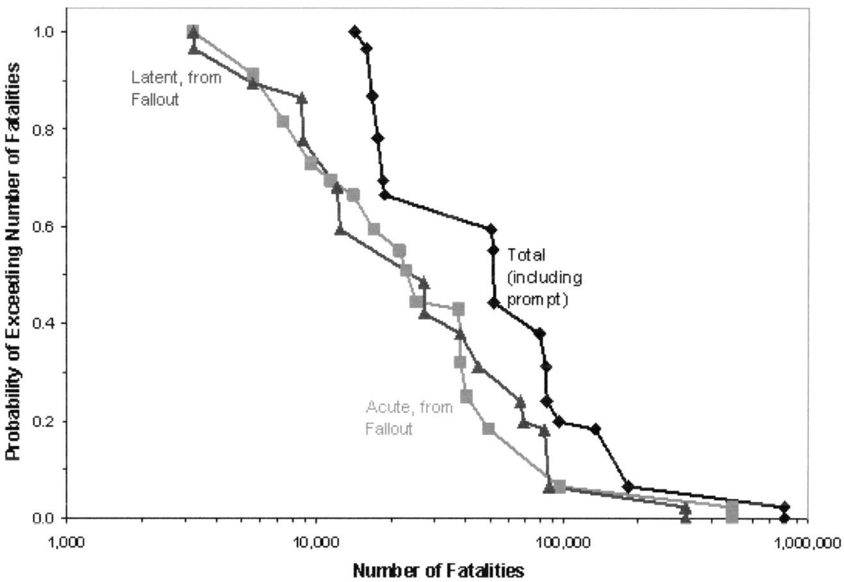

FIGURE 6.10(b) The probability of exceeding a given number of deaths due to acute and latent effects from external exposure to gamma-radiation fallout from a 300 kiloton earth-penetrator weapon at 3 meters' depth of burst on notional target B, assuming that the population is in the open. SOURCE: Estimates prepared for the committee by the Lawrence Livermore National Laboratory.

effects of fallout; 50,000 to 160,000 for deaths from latent effects from fallout; and 60,000 to 900,000 for total fatalities. For Target B, the corresponding intervals are 9,000 to 40,000 for deaths from acute effects of fallout; 10,000 to 60,000 for deaths from latent effects from fallout; and 20,000 to 90,000 for total fatalities.

Although the committee has not done a comprehensive analysis of the effect of wind direction for a wide range of yields, it is apparent that the casualty-reduction factor (the ratio of number of casualties for a surface burst to that for an EPW with a yield 25 times smaller) could be considerably lower or higher than the mean ratios given in Figure 6.8, depending on wind direction. For example, Figures 6.11(a) and (b) give total fatalities for a 10 kiloton EPW and a 250 kiloton surface burst for Targets A and B, respectively. Total fatalities are 10 to 40 times higher for the surface burst for Target A, depending on wind direction, with a mean 20 times higher. For Target B, the fatality ratio varies from 4 to 40, with a mean of 16; for comparison, the mean casualty ratio given in Figure 6.8 (from DTRA's HPAC code) is 18.

The model runs show significant fatalities from both an EPW and a surface-burst weapon. The numbers are larger when the attack is near a population center and if a wind that would blow the fallout into the population center is introduced in the calculations. Figures 6.11(a) and (b) show that, for a given wind direction, the estimated number of fatalities is significantly smaller for the lower-yield EPW. It is also worth noting, however, that with unfavorable winds the lower-yield EPW would cause about as many deaths as would the higher-yield surface burst with favorable winds. For example, 40,000 deaths result from attacks on Target A from the 10 kiloton EPW with the wind blowing from the west and the 250 kiloton surface burst with the wind blowing from the east. Similarly, 15,000 deaths result from attacks on Target B from the 10 kiloton EPW with the wind blowing from the southeast and the 250 kiloton surface burst with the wind blowing from the northwest. These numbers suggest that wind direction can be as important as a 25-fold difference in yield in determining civilian casualties from attacks in which fallout is the primary health hazard.[16] However, Figures 6.11(a) and (b) also show that for the same wind direction, with few exceptions, the number of fatalities from the surface burst are significantly larger than the number from the EPW. These comparisons indicate the sensitivity to wind of collateral damage to populations. However, an unfavorable wind for an EPW is, of course, also an unfavorable wind for a surface burst; the same is true for favorable winds. A population center downwind of either weapon is an unfavorable situation.

Fallout Exposure Pathways Not Considered

As noted above, the estimates produced by DTRA and LLNL of the numbers of deaths and injuries due to fallout include only the external gamma-ray dose from the deposition of fallout particles on ground surfaces.[17] These estimates do not include external doses of radiation from the passing cloud or internal doses of radiation from the inhalation of contaminated air or ingestion of contaminated food or water. The contribution of these exposure pathways to the acute radiation dose usually is not substantial and would not significantly alter the estimates presented above. Under some conditions, however, the contribution of other exposure pathways to the risk of latent cancer could be significant. Here the contribution of these other exposure pathways is reviewed in a semiquantitative manner.

External Dose from the Passing Cloud

For underground, surface, or near-surface nuclear explosions, the radioactive fallout is mixed with a large mass of ejecta in the main cloud or base surge. These clouds are dense, and most of the mass at

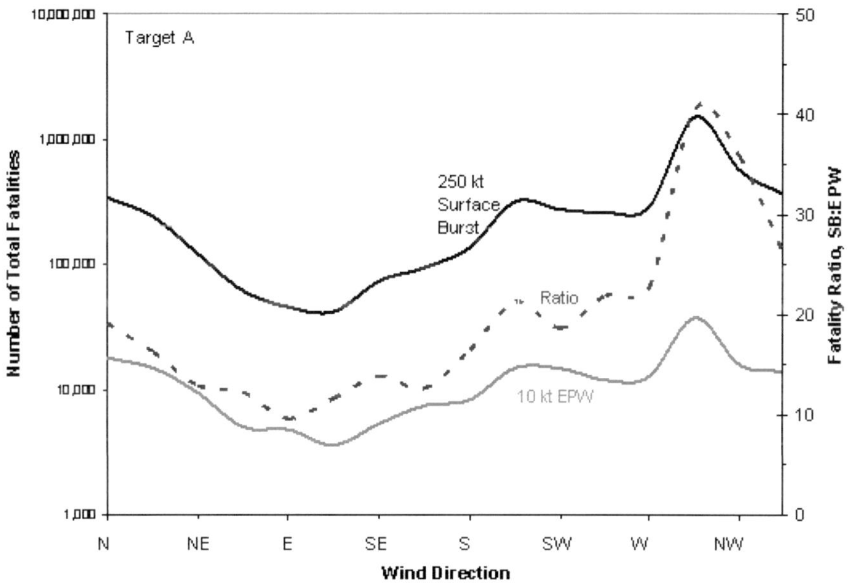

FIGURE 6.11(a) The estimated number of total fatalities (deaths due to prompt effects, plus acute and latent effects from fallout) from attacks with a 10 kiloton earth-penetrator weapon (EPW) at 3 meters' depth of burst and a 250 kiloton surface burst on notional target A (30 kilometers northwest of target A). SOURCE: Estimates prepared for the committee by the Lawrence Livermore National Laboratory.

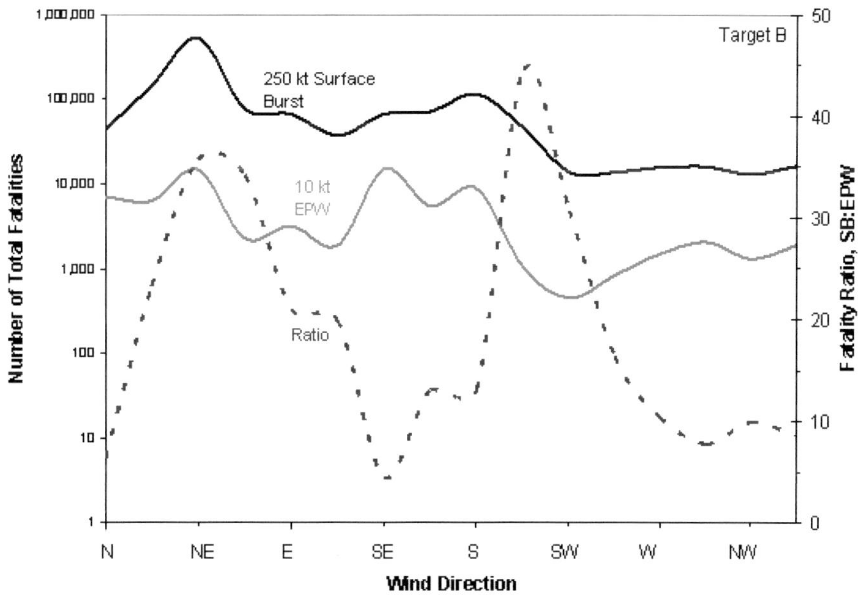

FIGURE 6.11(b) The estimated number of total fatalities (deaths due to prompt effects plus acute and latent effects from fallout) from attacks with a 10 kiloton earth-penetrator weapon (EPW) at 3 meters' depth of burst and a 250 kiloton surface burst on notional target B. SOURCE: Estimates prepared for the committee by the Lawrence Livermore National Laboratory.

close-in locations consists of large particles that deposit rapidly. Thus, most of the external dose received by persons within several kilometers of the detonation point is due to radiation from the deposited material rather than from the airborne cloud itself. Also, at close-in distances, cloud passage occurs during a rather short period of time; this is another reason that the integrated exposure from cloud passage tends to be small relative to the long-term exposure from radionuclides deposited on the ground.

During the 1950s when atmospheric nuclear testing was conducted at the Nevada Test Site (NTS), there were a number of sets of measurements of the rate of exposure before, during, and after the passage of clouds from a variety of types of nuclear tests.[18] In most cases there was no measurable exposure rate that could be attributed to exposure to the cloud itself—at least not in comparison with the exposure rate derived from exposure to material on the ground. In general, the radiation dose received from the passage of the cloud itself is not a significant fraction of the dose received as a result of total external exposure.

Inhalation of Contaminated Air

In addition to external exposure, individuals may also be exposed to radiation by inhalation of fallout particles, either during the passage of the cloud or subsequently owing to resuspension of deposited particles by wind, plowing, vehicle travel, or other disturbances of the surface. Based on measured external gamma-radiation exposure rates and air concentrations observed downwind of explosions at the NTS, the whole-body inhalation dose was calculated to have ranged for most organs from 1 to 20 percent of the dose that resulted from the ingestion of contaminated food.[19] However, the relative dose to the organs of the gastrointestinal tract via inhalation can be much larger, up to 80 percent of the dose from ingestion. This larger dose is due to the entrance during cloud passage of large particles into the upper respiratory tract, from which the particles are coughed up and swallowed.

The inhalation of resuspended radionuclides is a pathway of interest under only a few special circumstances—primarily with respect to the inhalation of radionuclides that do not cross biological barriers easily but can be retained over very long periods if inhaled. The most notable example of such a radionuclide is plutonium. If a nuclear device performs correctly, plutonium has not been found to be a significant source of radiation dose.

In general, inhalation is not very significant compared with other pathways of exposure. Consideration of this pathway would not significantly increase the casualty estimates presented above.

Ingestion of Contaminated Food and Water

The consumption of contaminated water has not been found to be a significant exposure pathway following nuclear tests at the NTS. Although deposition on water surfaces does occur, it has not been a significant source of exposure because dilution is rapid for persons living downwind of the NTS. The aquatic pathway was of greater concern following the Chernobyl accident, which contaminated one of the watersheds supplying water to the Kyiv Reservoir. Even in this situation, however, the consumption of contaminated water was not a substantial pathway.

Contamination of some lake systems following the Chernobyl accident in locations as far away as Sweden and Norway was more of a problem for lakes having a large surface area, shallow depth, and limited inflow and outflow. In this case the direct consumption of water was not of interest; rather, the fish in such locations were found to have elevated levels of cesium-137. It is doubtful that a similar situation would occur following a nuclear explosion, as the amounts of long-lived radionuclides created in an explosion are much smaller than are the amounts produced in a reactor that has operated for several years.

The consumption of food contaminated by fallout from a nuclear test, however, has proven to be a major problem both at the NTS[20,21] and the Semipalatinsk Polygon, a nuclear test site in the Soviet Union.[22] The nature of this problem was not fully appreciated until 1963—at about the time that atmospheric testing by the United States and the former Soviet Union was ending. By far the largest concern has been associated with iodine-131, which has a half-life of 8 days. It is by the combination of several fairly unique circumstances that this radionuclide has been the major radionuclide of concern from the viewpoint of food contamination for both nuclear weapons tests and for reactor accidents.

Substantial amounts of ^{131}I activity are created by nuclear explosions; this radionuclide is also volatile and does not condense on particles until late, at which time it becomes associated with the surfaces of fallout particles.[23] Most of the total surface activity is contained on the smaller particles, so ^{131}I is typically transported farther. The smaller particles are also preferentially retained by vegetation,[24] from which they are lost with a half-retention time of about 10 days. A milk cow, if it is receiving its full quota of food from fresh pasture, will consume per day the amount of ^{131}I that is contained on about 50 square meters,[25] and it will secrete up to 1 percent of that daily intake into a liter of milk.[26] Typically, a human consuming milk will concentrate 30 percent of his or her intake into the thyroid gland. The thyroid is a very small gland, weighing about 20 grams in adults and only about 2 grams in infants. Thus, iodine is preferentially retained on vegetation, which the cow efficiently samples and rapidly secretes into milk; an infant then concentrates a large fraction of that iodine in milk into an extremely small gland, thus producing a relatively large dose.

"Backyard cows" are of more concern, as such cows typically consume more pasture as opposed to stored feed, and the owners frequently drink more than an average amount of milk. Goats are also of more concern; they graze less territory, but they secrete about 10 times more of their daily intake of iodine into 1 liter of milk. For nuclear explosions outside the United States, the consumption of milk from other animals, such as sheep, horses, and camels, should be considered. The milk-transfer factors for these animals are not well known. Scientists at the National Cancer Institute are conducting a research program to determine such factors, but the results are not yet published.[27] Frequently, the milk from such animals is not consumed immediately but is made into other products, thus providing some opportunity for the ^{131}I to decay before being consumed.

For a hypothetical device (with approximately 50 percent fission fraction, i.e., 50 percent of the explosive power from fusion) that produces an integrated external dose of 1 rad, the dose to an infant's thyroid would be about 16 rads from the consumption of milk with ^{131}I and a few other radionuclides (^{132}Te, ^{132}I, ^{133}I, and ^{135}I). These results are scaled from published calculations made for NTS shots.[28]

Other radionuclides of concern in terms of contaminated foods are ^{89}Sr, ^{90}Sr, and ^{137}Cs. These share the characteristics of high fission yield (the fraction of fissions that produce the radionuclide or its precursors), volatility (of the radionuclide or its precursors), and efficient secretion into milk. Other organs of concern are the digestive tract, red bone marrow, and bone surfaces.

So far, it has been assumed for this discussion that the persons and the milk animals are collocated. This is frequently not the case. Reconstruction of thyroid dose from past events has included elaborate attempts to reconstruct sources of milk or movement of milk from one region to another.[29,30] If this type of predictive assessment were to be included in an analysis of effects, it would be necessary to have a database that gave the population density of humans as well as milk animals.

It is important to note that this pathway, consumption of contaminated food, can be relatively more important for fallout from nuclear explosion accidents in nonurban areas in the sense that milk animals are more likely to be located in rural areas. The problem of contaminated milk supplies following a nuclear accident is now widely known, especially after the Chernobyl accident.[31] Thus, it is frequently possible to eliminate this pathway by a variety of means (taking animals off pasture, discarding the milk, blocking the uptake of iodine by the human thyroid by feeding large quantities of stable iodine, use of cesium binders, and so on). Such elimination of this pathway would require that local inhabitants were adequately warned; that sufficient monitoring devices, iodine supplies, and distribution systems were available; and that alternate food supplies were available.

Contamination of other types of food crops would also occur. After milk, the food of most concern is fresh, leafy vegetables. Such vegetables are efficient in capturing fallout and are typically consumed fresh on a daily basis during the growing season. This practice provides an opportunity for a direct and rapid pathway to humans following deposition of fallout but, again, this pathway can be eliminated by an informed population with an adequate infrastructure. Other types of food crops typically have less ability to capture fallout or have more indirect and longer pathways to humans. The longer pathways allow for both radioactive decay and the loss of retained material from the crops. Pathways of possible concern include the consumption of meat from grazing animals, poultry, and eggs. Grain crops are not usually of concern unless they are harvested immediately after deposition of fallout.

The consumption of contaminated food is unlikely to result in any acute health effects, but it could in some circumstances increase significantly the number of latent cancers that would be expected in the affected population. An accurate estimate of the number of latent cancer fatalities from this exposure pathway would require estimating the amount of contamination in milk and various other foods, the consumption of these foods by the population, the internal dose from each radionuclide to each organ, and the use of organ-specific risk coefficients.[32]

Exposure to Fallout at Very Great Distances

The computer codes used for this study do not consider deposition at very great distances. If clouds are lofted to substantial heights and later encounter precipitation systems, there can be areas of enhanced deposition very far away. Such an event occurred in the area of Troy, New York, following the NTS test Simon in April 1953.[33] An area of enhanced deposition also occurred in Indiana following the Trinity test in New Mexico in 1945. This deposition was eventually detected only after contaminated straw used in the packing of x-ray film was noted to have exposed the film.[34]

Of more recent interest were the areas of enhanced deposition that resulted from the Chernobyl accident. Contamination was sufficiently high in areas of several countries far from the accident (e.g., Sweden, Norway, and the United Kingdom) that restrictions on food use were implemented by national authorities.[35] The occurrence of such areas is difficult to predict and, depending on a country's resources, may go undetected.

"Global fallout" is a general term that describes the injection of nuclear debris into the stratosphere. Such fallout returns to Earth slowly, and with a half-time of about 1 year, most of the short-lived radionuclides would have decayed before the fallout returned to Earth. It takes a large explosion to produce such injections, on the order of hundreds of kilotons. Much of the experience with global fallout resulted from the large tests conducted by the United States and the Soviet Union from 1961 to 1963, although earlier large tests in 1952, 1954, 1956, and 1958 also produced global fallout. Concern was largely focused on ^{90}Sr and ^{137}Cs, each of which has a half-life of about 30 years. Under unusual circumstances, such as the large-scale subsidence of air masses or the penetration of large thunderstorms into the stratosphere, the deposition of ^{131}I was also noted.[36] The negative worldwide reaction to global fallout was intense in the early 1960s, and this was one of the more important factors that resulted in the agreement to stop atmospheric tests by the United States, the United Kingdom, and the Soviet Union.

HEALTH EFFECTS FROM ATTACKS ON FACILITIES FOR STORING AND PRODUCING NUCLEAR WEAPONS AND RADIOACTIVE MATERIAL

The health effects resulting from attacks with conventional weapons on nuclear-weapon storage facilities depend on the detailed design of the nuclear weapons being attacked. Because the design details of enemy nuclear weapons are unknown (and could not be discussed in this document in any case), the committee cannot provide quantitative estimates.

However, if the enemy nuclear weapons are "one-point safe" (i.e., there is less than 1 chance in a million that the yield will be over 4 pounds when the high explosive is initiated and detonated at a single point), then the main risk to nearby civilian populations would result from the dispersal of radioactive material. The greatest such risks would arise from weapons containing plutonium. Even in this case, however, the dispersal of plutonium from tens of weapons would be unlikely to cause deaths or acute illnesses in civilian populations. Dispersal of plutonium could, however, result in thousands of latent cancer deaths if kilogram quantities of plutonium aerosol were dispersed in densely populated areas.[37] If the weapons contain no plutonium (only highly enriched uranium), then this concern would be very much reduced. If an enemy's nuclear weapons are not one-point safe, it is possible that a conventional attack could result in a nuclear detonation. In this case, the effects on nearby civilian population would be similar to those estimated in Figure 6.7.

If the enemy's nuclear weapons are not certifiably one-point safe, then the assessment of possible yields is much more complicated. It seems probable that even an early-stage nuclear country or group would desire some degree of safety in order to preserve both the weapon and the nuclear material for the use for which it was intended. Consequently, other techniques will likely be employed to create safe operating conditions for the weapons. For example, the weapon components can be kept in separate locations, ready to be assembled quickly for possible use (as was done with a number of U.S. weapons). There may also be mechanical safety devices in place that lead to a low probability of unintentional detonation (even if not as quantitative as the one-point safety criterion). And many possible weapons will be "partially safe"—i.e., their one-point yield will be much less than their design yield. As a result, the probability of significant nuclear yield from a conventional attack is quite low—but cannot be completely ruled out. In that case an upper limit for the effects is similar to the limits estimated in Figure 6.7. However, the most likely outcome of such an attack is dispersal of the nuclear material, the equivalent of the dirty bomb scenario discussed below.

The dispersal of radioactive materials from a non-nuclear explosion would be possible, for example, if sympathetic detonation of high explosives led to dispersal of the radioactive material either in weapons or in a facility such as a reprocessing plant. In this case, the effects would be similar to those discussed for what have been called "radiological" weapons.

The term "radiological weapon" is extremely broad and imprecise. A radiological weapon could involve a device using any of hundreds of radionuclides, in quantities ranging from harmless to lethal, in physical and chemical forms that are easy or impossible to disperse efficiently.

Calculations done by others[38] indicate that the acute effects of a "dirty bomb" containing even a potent radioactive source would in most cases not extend beyond the lethal radius of the high explosive used to disperse the radioactive material. Accordingly, the committee expects that a conventional attack on a facility containing radiological weapons or radioactive materials would be unlikely to produce a substantial number of civilian deaths or acute illnesses, beyond those caused directly by the conventional attack itself. The number of latent cancer deaths that might result from a dispersal of radioactive material would depend sensitively on the type and amount of material dispersed (as well as the density of nearby civilian populations and whether these populations were evacuated from the area after the attack). Uncertainties in the source term make quantitative estimates impossible, but the estimates given above for plutonium dispersal indicate the consequences of the dispersal of a very large mass of highly radioactive material.

ENVIRONMENTAL EFFECTS

In addition to the health effects mentioned above, a variety of environmental effects can be expected from nuclear explosions near the ground's surface. Following the explosion of a nuclear weapon, the fallout area is intensely radioactive. However, as noted above, the rate of external exposure to gamma radiation decreases rapidly with time, and the denial of land use due to fallout is not of great concern relative to other effects of fallout. This is in marked distinction to the situation to be expected following a major reactor accident such as that at Chernobyl,[39] because of the much greater releases of long-lived ^{137}Cs. Denial of the use of water would be expected to be of even less concern, except under very unusual circumstances, because of the very rapid dilution of fallout deposited on surface waters. It is unlikely that significant contamination of groundwater would occur, except in areas immediately adjacent to an explosion of an earth-penetrator weapon. Although underground facilities could be built below the water table and kept dry by diversion and pumping, most facilities are expected to be above the water table. The groundwater in the immediate area of an underground burst would be contaminated, but the greatest release of radioactivity would be from activated materials that are spread onto the surface. A bunker facility is highly unlikely to be built in groundwater. Groundwater is likely to be in the fallout area. However, the greatest release of radioactivity would be from activated material that is spread onto the water surface. The transport of radionuclides due to the movement of groundwater will be difficult to evaluate with any useful certainty as it is a very site-specific phenomenon. U.S. experience at the Nevada Test Site indicates that the movement of radionuclides by groundwater is quite limited, although some radionuclides have been found off-site after many decades. Such effects of groundwater will be far less than the effects of blast, fire, and on-the-ground fallout.

The radiation sensitivity of all other mammals is generally about the same as that for humans. Thus, in areas where humans are killed or injured by radiation, the same lethality for animals would be expected. If large herds of farm animals were affected, poor sanitation could become a significant problem.

Plant species have a broad range of sensitivity to radiation.[40] Among the more sensitive are some species of trees, particularly pine and spruce, which are roughly as sensitive as humans are. Thus, it is conceivable that forests could be killed, which in turn could result in forest fires. The demise of the pine forest near the Chernobyl plant was one notable example of this effect.[41] Other vegetation types might also die as a result of contamination within the range of concern for human lethality. It is not likely that effects in excess of that indicated for pine forests would occur.

Recently, there has been a focus on evaluating the possible effects of radiation on other members of an ecological system.[42] Generally, concern is limited to the possible effects on populations of species rather than on individual members of an ecosystem. It is not expected that effects other than those mentioned above would be of significance.

In the past there has been concern that large numbers of nuclear explosions might lead to large-scale disruption of the environment, including depletion of stratospheric ozone due to nitrogen oxides produced by the fireball, and changes in climate due to the soot and other aerosols released from burning cities. These concerns are relevant only with the detonation of thousands of high-yield weapons. No significant environmental disruptions would be expected to occur beyond the areas directly affected by the prompt effects from one or a few nuclear explosions and the fallout that, depending on the amount of soil entrained and the fission fraction of the weapon(s), can persist at dangerous levels for at least a year.

EFFECTS OF ATTACKS ON CHEMICAL AND BIOLOGICAL WEAPONS FACILITIES

Earth-penetrator weapons—conventional or nuclear—provide a means to defeat or destroy hardened and/or deeply buried facilities used for the production or storage of chemical and biological agents. In this context, there are three important questions:

1. To what extent can conventional or nuclear weapons destroy such facilities or the chemical and biological agents that they contain?
2. To what extent would a conventional or nuclear attack on such a facility result in the release of chemical and biological agents?
3. If chemical or biological agents are released as a result of an attack, what would be the health consequences for the nearby civilian population?

The answers to the first two questions depend critically on detailed information about the facility, including its location, construction, and layout; the type and number of agent containers and their placement within the facility; and the amount and type of agent and the form in which it is stored. This information would provide the basis for targeting, selecting weapons, and estimating how much agent might be destroyed or released. Unfortunately, detailed information of this kind is likely to be highly uncertain or unavailable for many potential targets. Existing estimates of the amount of agent that might be destroyed or dispersed in a nuclear attack are based entirely on computer models using greatly simplified assumptions. In the case of conventional attacks, experiments also have been conducted using prototype facilities, with surrogates in place of live agent. Even with all of these qualifications, certain important points can be made:

- It is important to distinguish between the defeat or destruction of a chemical or biological weapons facility and the destruction of the chemical or biological agent contained within it. Facilities can be defeated or destroyed without destroying the agent inside. For example, a nuclear EPW could crush a storage facility under 100 meters of rock without destroying (or releasing) any agent. Similarly, conventional weapons could collapse surface or near-surface entrances to such a facility and thereby hinder or delay the use of agents by the enemy.
- If facilities or storage areas are penetrated by a nuclear or conventional weapon, significant degradation (thereby reducing potential releases) can be effected by heat (>1,000 degrees Fahrenheit and residence time >20 to 30 seconds).
- The thermal destruction of chemical or biological agents requires the deposition of large amounts of heat throughout the agent. Although existing conventional earth-penetrator weapons, such as the GBU-24, can penetrate and destroy shallow buried facilities, they cannot deliver enough energy to reliably and completely destroy large stockpiles of chemical or biological agent, although they may substantially degrade the agents. Non-nuclear agent-defeat weapons now under development may ultimately prove to be more effective. However, the BLU-118B thermobaric bomb, if detonated within the chamber, may be able to destroy the agents.[43]
- Nuclear weapons are capable of delivering the very large amounts of heat and radiation required to destroy large stocks of chemical and biological agents. In order for this heat and radiation to be deposited throughout the agent, the nuclear weapon must be detonated in the chamber where the agent is stored. Weapons detonated several meters above, below, or to the side of storage facilities may be much less effective in destroying the agent.
- The manner in which the agent is stored (e.g., the types of containers, location in multiple storage

rooms) and the proximity of the detonation of the weapon may result in significant variations in the radiation doses and thermal histories of the agent in different parts of a facility. Given the many unknowns, a conservative analysis must presume some release of the agent in a viable form if the facility is breached, regardless of the type of weapon used.

- The amount of agent that likely would be released is extremely difficult to estimate accurately. DTRA estimates that an attack with existing conventional weapons could cause the release in respirable form of 0.1 to 5 percent of the agent inventory.[44] Calculations indicate that an attack with a nuclear weapon could result in comparable releases if the weapon was detonated close to but not within a facility, but much smaller releases if the weapon was detonated in the same room as the agent.

The consequences of a release of agent can be estimated using computer codes that model the dispersion of agent and subsequent human exposures and health effects. Although these models are similar in some ways to those used to estimate the consequences of nuclear fallout, the transport of chemical and biological agents is more complicated and more uncertain. The particle size distribution of biological agents and some chemical agents may change during transport. The persistence of both chemical and biological agents depends on temperature, humidity, exposure to ultraviolet light, precipitation, and agent-surface reactions. The importance of these factors differs for each type of agent, but for most chemical and biological agents of concern, one may expect a rapid degradation in their toxicity or viability within hours to days—minutes in the case of some biological agents—following a release into the open air. In contrast, some agents, such as anthrax spores, mustard, and lewisite, may persist for many years.

Although many studies have validated and verified the fate of chemical agents during transport, few are available for biological agents, and the fate of biological agents during transport is therefore difficult to model. In addition, the dose-response relationships are uncertain for many biological agents and often are very sensitive to the age and health status of the person exposed. As with fallout, the estimated number of casualties resulting from a given release of agent can vary across a very wide range, depending on weather conditions, the density and distribution of the population, the proportion of the population that is sheltered and/or equipped with protective gear, and the availability of prompt medical care.

Though multiple experiments using biological and chemical agent surrogates have been conducted, they provide an imperfect database. Actual experience that might be used to validate models is limited to one release of biological agent (anthrax spores) at Sverdlovsk in 1979 and one release of chemical agent (sarin) in the Tokyo subway system in 1995. Media reports of the use of chemical agents by the Iraqi government against Kurdish villages do not provide sufficient information about agent concentrations or delivery method to be useful, and the case of the letters containing anthrax sent through the U.S. Postal Service in 2001 is of limited relevance to the type of situation considered here.

At the request of the committee, DTRA estimated the average number of fatalities that would result from various releases of sarin (a nerve agent) and anthrax at three locations in the Washington, D.C., area: the city center and 10 and 50 kilometers northwest of the city center. In each case, releases of 1 to 10,000 kilograms of sarin and 1 gram to 10 kilograms of weaponized dry anthrax spores were considered, corresponding to releases of 0.001 to 10 percent of an inventory of 100 tons of sarin and 100 kilograms of anthrax. The average number of fatalities from prompt and acute effects of fallout resulting from attacks with nuclear EPWs with yields of 3 and 30 kilotons were also estimated. The population was assumed to be static and entirely in the open with no protection. The results are shown in Figures 6.12(a) through 6.12(c).

The estimated mean number of fatalities resulting from a 1,000 kilogram release of sarin (1 percent of a 100 ton inventory) ranges from about 100 to 1,000 depending on the location of the release. In

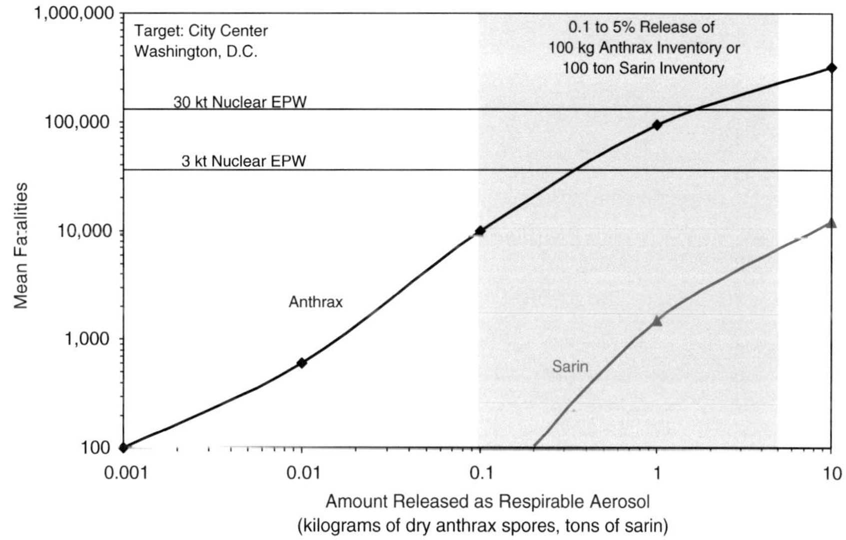

FIGURE 6.12(a) Illustrative example: Estimated mean number of fatalities from releases of sarin or anthrax at city center of Washington, D.C., compared with the mean number of fatalities resulting from 3 kiloton and 30 kiloton nuclear earth-penetrator weapon (EPW) explosions at the same location. SOURCE: Estimates prepared for the committee by the Defense Threat Reduction Agency.

contrast, the mean number of fatalities from a release of 1 kilogram of dry anthrax spores (1 percent of a 100 kilogram inventory) is about 100,000 for each of the three locations. For the reasons previously discussed (i.e., the generally fragile nature of most toxic agents), the calculated number of fatalities for the release of sarin is certainly too high.

For comparison, the estimated mean number of fatalities ranges from 7,000 to 40,000 for a 3 kiloton EPW, and from 30,000 to 130,000 for a 30 kiloton EPW, depending on the location. Because the expected number of fatalities from a relatively low yield (3 kiloton) nuclear EPW exceeds that from an extremely large (10,000 kilogram) release of sarin, it is highly unlikely that a nuclear attack would result in smaller total collateral effects than those from a conventional attack against a facility for the storage or production of chemical agents. In contrast, releases of as little as 0.1 kilogram of anthrax result in a calculated number of fatalities that is comparable to that estimated for a 3 kiloton nuclear EPW.

NOTES

1. International Commission on Radiological Protection. 1991. "Recommendations of the International Commission on Radiological Protection," *Annals of the ICRP*, ICRP Publication 60, Vol. 21 (1-3), Pergamon Press, Oxford.

2. For example, if 100 people received an average effective dose of 1 sievert, 5 would be expected to die from cancer as a result of this exposure. The "effective" dose is the sum of equivalent doses to various organs multiplied by a weighting factor that is established according to the estimated likelihood of a cancer occurring in that organ; the sum of all weighting factors is 1. The effective dose is roughly equal to the whole-body dose for external exposure to gamma rays.

3. M. Otake and W.J. Shull. 1984. "In Utero Exposure to A-bomb Radiation and Mental Retardation: A Reassessment," *Br. J. Radiol.*, Vol. 57, pp. 409-414.

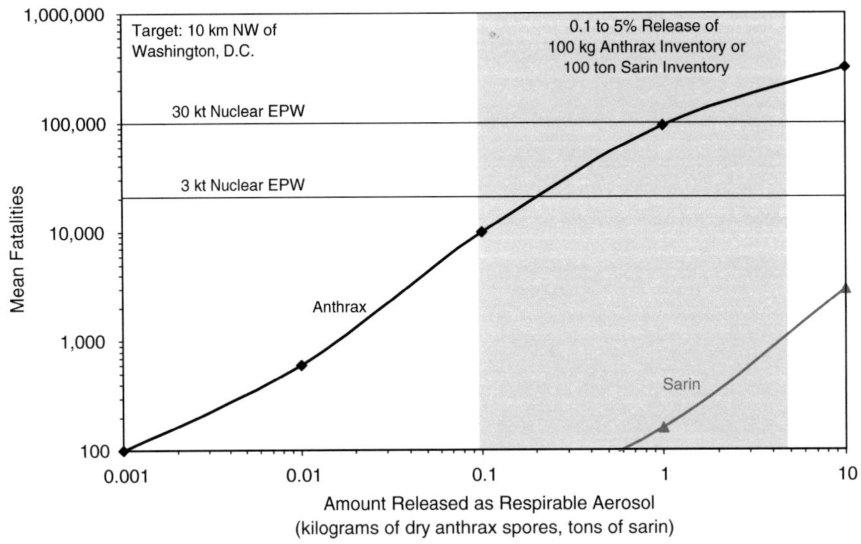

FIGURE 6.12(b) Illustrative example: Estimated mean number of fatalities from releases of sarin or anthrax 10 kilometers northwest of Washington, D.C., compared with the mean number of fatalities resulting from 3 kiloton and 30 kiloton nuclear earth-penetrator weapon (EPW) explosions at the same location. SOURCE: Estimates prepared for the committee by the Defense Threat Reduction Agency.

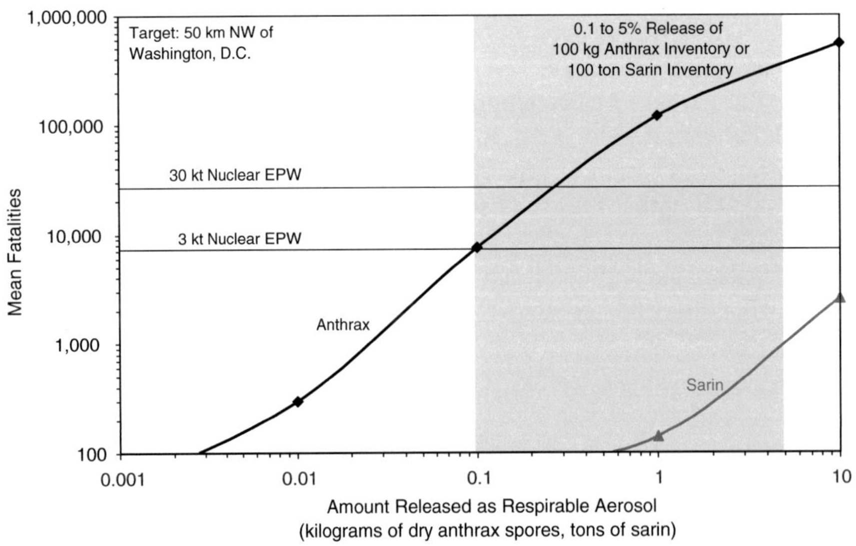

FIGURE 6.12(c) Illustrative example: Estimated mean number of fatalities from releases of sarin or anthrax 50 kilometers northwest of Washington, D.C., compared with the mean number of fatalities resulting from 3 kiloton and 30 kiloton nuclear earth-penetrator weapon (EPW) explosions at the same location. SOURCE: Estimates prepared for the committee by the Defense Threat Reduction Agency.

4. International Commission on Radiological Protection. 2001. "Doses to the Embryo and Fetus from Intakes of Radionuclides by the Mother," *Annals of the ICRP,* ICRP Publication 88, Vol. 31 (1-3), Pergamon Press, Oxford.

5. A.K. Junk, Y. Kundiev, P. Vitte, and B.V. Worgul. 1999. *Ocular Radiation Risk Assessment in Populations Exposed to Environmental Radiation Contamination*, Kluwer Academic Publishers, Boston, Mass.

6. M. Yamada, F.L. Wong, S. Fujiwara, M. Akahoshi, and G. Suzuki. 2004. "Noncancer Disease Incidence in Atomic Bomb Survivors, 1958-1998," *Radiat. Res.*, Vol. 161, pp. 622-632.

7. The yield options for the proposed RNEP cover a wide range, and the committee reviewed analyses by DTRA and others that covered a wide range of EPW yields. Although the hardest and deepest targets require EPW yields of 300 to 1,000 kt, other targets of interest could be destroyed with EPWs with yields of 1 to 10 kt. The committee therefore did a parameter analysis in which the EPW yield ranged from 1 to 1,000 kt. The mean number of casualties over this range of EPW yields is shown in Figures 6.6 and 6.7.

8. A "serious injury" is defined as one requiring hospitalization for 60 days or more.

9. This is true for entry times of less than 6 months.

10. National Council on Radiation Protection and Measurement. 1993. *Limitation of Exposure to Ionizing Radiation,* Report No. 116, Washington, D.C.

11. Institute of Medicine. 1997. *An Evaluation of Radiation Exposure Guidance for Military Operations,* Interim Report, National Academy Press, Washington, D.C.; Institute of Medicine. 1999. *Potential Radiation Exposure in Military Operations,* National Academy Press, Washington, D.C.

12. The effects discussed here are only those from the nuclear weapon and its direct effects and do not include chemical and biological agents, which are discussed in the section of this chapter entitled "Effects of Attacks on Chemical and Biological Weapons Facilities."

13. If the depth of burst is increased from 3 meters to 10 meters, total casualties increase only by about 10 percent at high yields to 20 percent at low yields, owing largely to the increase in the amount of soil excavated by the explosion.

14. Samuel Glasstone and Philip J. Dolan (eds.). 1977. *The Effects of Nuclear Weapons,* U.S. Government Printing Office, Washington, D.C.

15. Using a risk coefficient of 0.05 per sievert, the National Atmosphere Release Advisory Center estimated cancer deaths for populations receiving doses greater than 1 millisievert (100 millirems), which is roughly equal to the average annual dose due to external radiation from cosmic rays and radionuclides in soil. This threshold was used only to limit the complexity of the calculation; the committee takes no position on whether a threshold exists in the dose-response relationship.

16. The National Command Authority and the deployers have opportunities and the responsibility to execute an attack on HDBTs in ways to minimize collateral damage by taking into account wind direction as well as yield.

17. Lethal beta skin burns, the major cause of fatality from acute effects of fallout at Chernobyl, are not considered. KDFOC does not consider beta burns in its analyses because burns are not considered a first-order lethality effect, like prompt and local fallout. For residual effects, it considers only whole-body gamma groundshine from fallout particles greater than 5 microns. Beta burns from such fallout particles would not be acutely lethal except in areas where gamma radiation would already have been lethal, thus, double-counting. HPAC does not include beta-induced injuries—all casualties are derived from effects of gamma radiation. The main problem with beta injuries is that the material must come into contact with skin, and HPAC has no means to determine the orientation and skin exposure posture of the population, nor the secondary beta burns received by people touching a surface contaminated with beta particles. Secondary beta burns are potentially a problem, but there is no way to determine casualties because the total population is not affected.

18. H.G. Hicks. 1990. "Additional Calculations of Radionuclide Production Following Nuclear Explosions and Pu Isotopic Ratios for Nevada Test Site Events," *Health Phys.,* Vol. 59, pp. 515-523.

19. Y.C. Ng, L.R. Anspaugh, and R.T. Cederwall. 1990. "ORERP Internal Dose Estimates for Individuals," *Health Phys.,* Vol. 59, pp. 693-713.

20. Y.C. Ng, L.R. Anspaugh, and R.T. Cederwall. 1990. "ORERP Internal Dose Estimates for Individuals," *Health Phys.,* Vol. 59, pp. 693-713.

21. F.W. Whicker and T.B. Kirchner. 1987. "PATHWAY: A Dynamic Food-Chain Model to Predict Radionuclide Ingestion After Fallout Deposition, *Health Phys.,* Vol. 52, pp. 717-737; F.W. Whicker, T.B. Kirchner, L.R. Anspaugh, and Y.C. Ng. 1996. "Ingestion of Nevada Test Site Fallout: Internal Dose Estimates," *Health Phys.,* Vol. 71, pp. 477-486.

22. K. Gordeev, I. Vasilenko, A. Lebedev, A. Bouville, N. Luckyanov, S.L. Simon, Y. Stepanov, S. Shinkarev, and L. Anspaugh. 2002. "Fallout from Nuclear Tests: Dosimetry in Kazakhstan," *Radial. Environ. Biophys.,* Vol. 41, pp. 61-67.

23. H.G. Hicks. 1982. "Calculation of the Concentration of Any Radionuclide Deposited on the Ground by Off-site Fallout from a Nuclear Detonation, *Health Phys.,* Vol. 42, pp. 585-600.

24. L.R. Anspaugh, S.L. Simon, K.I. Gordeev, I.A. Likhtarev, R.M. Maxwell, and S.M. Shinkarev. 2002. "Movement of Radionuclides in Terrestrial Ecosystems by Physical Processes," *Health Phys.*, Vol. 82, pp. 669-679.

25. J.J. Koranda. 1965. *Agricultural Factors Affecting the Daily Intake of Fresh Fallout by Dairy Cows,* UCRL 12479, Lawrence Livermore National Laboratory, Livermore, Calif.

26. National Cancer Institute. 1997. *Estimated Exposures and Thyroid Doses Received by the American People from Iodine-131 Following Nevada Atmospheric Nuclear Bomb Tests: A Report from the National Cancer Institute*, U.S. Department of Health and Human Services, Washington, D.C.

27. Steven L. Simon, National Cancer Institute, Bethesda, Md., personal communication, 2003.

28. Y.C. Ng, L.R. Anspaugh, and R.T. Cederwall. 1990. "ORERP Internal Dose Estimates for Individuals," *Health Phys.*, Vol. 59, pp. 693-713.

29. R.A. Kerber, J.E. Till, S. Simon, J.L. Lyon, D.C. Thomas, S. Preston-Martin, M.L. Rallison, R.D. Lloyd, and W. Stevens. 1993. "A Cohort Study of Thyroid Disease in Relation to Fallout from Nuclear Weapons Testing," *JAMA,* Vol. 270, pp. 2076-2082.

30. National Cancer Institute. 1997. *Estimated Exposures and Thyroid Doses Received by the American People from Iodine-131 Following Nevada Atmospheric Nuclear Bomb Tests: A Report from the National Cancer Institute*, U.S. Department of Health and Human Services, Washington, D.C.

31. United Nations Scientific Committee on the Effects of Atomic Radiation (UNSCEAR), 1988, *Sources, Effects and Risks of Ionizing Radiation,* UNSCEAR 1988 report to the General Assembly, with annexes, United Nations, New York, Sales No. E.88.IX.7; also, United Nations Scientific Committee on the Effects of Atomic Radiation (UNSCEAR), 2000, *Sources, Effects and Risks of Ionizing Radiation,* UNSCEAR 2000 report to the General Assembly, with scientific annexes, United Nations, New York, Sales No. E.00.IX.4.

32. Such values are given in J.S. Puskin and C.B. Nelson, 1995, "Estimates of Radiogenic Cancer Risks," *Health Phys.*, Vol. 69, pp. 93-101.

33. H.M. Clark. 1954. "The Occurrence of an Unusually High-Level Radioactive Rainout in the Area of Troy, N.Y.," *Science,* Vol. 119, pp. 619-622.

34. J.H. Webb. 1949. "The Fogging of Photographic Film by Radioactive Contaminants in Cardboard Packaging Materials," *Phys. Rev.*, Vol. 76, pp. 375-380.

35. United Nations Scientific Committee on the Effects of Atomic Radiation (UNSCEAR), 1988, *Sources, Effects and Risks of Ionizing Radiation,* UNSCEAR 1988 report to the General Assembly, with annexes, United Nations, New York, Sales No. E.88.IX.7; also United Nations Scientific Committee on the Effects of Atomic Radiation (UNSCEAR), 2000, *Sources, Effects and Risks of Ionizing Radiation,* UNSCEAR 2000 report to the General Assembly, with scientific annexes, United Nations, New York, Sales No. E.00.IX.4.

36. L. Machta. 1963. "Meteorological Processes in the Transport of Weapon Radioiodine," *Health Phys.,* Vol. 9, pp. 1123-1132.

37. Steven A. Fetter and Frank von Hippel. 1990. "The Hazard from Plutonium Dispersal by Nuclear-Warhead Accidents," *Science and Global Security,* Vol. 2, pp. 21-41.

38. For example, see National Research Council, 2002, *Making the Nation Safer,* National Academies Press, Washington, D.C., p. 49.

39. United Nations Scientific Committee on the Effects of Atomic Radiation (UNSCEAR). 2000. *Sources, Effects and Risks of Ionizing Radiation,* UNSCEAR 2000 report to the General Assembly, with scientific annexes, United Nations, New York, Sales No. E.00.IX.4.

40. S. Ichikawa and A.H. Sparrow. 1967. *Radiat. Bot.,* Vol. 7, pp. 429-441.

41. United Nations Scientific Committee on the Effects of Atomic Radiation (UNSCEAR). 1996. *Sources and Effects of Ionizing Radiation*, UNSCEAR 1996 report to the General Assembly, with annex, United Nations, New York, Sales No. E.96.IX.3.

42. United Nations Scientific Committee on the Effects of Atomic Radiation (UNSCEAR). 1996. *Sources and Effects of Iionizing Radiation,* UNSCEAR 1996 report to the General Assembly, with annex, United Nations, New York, Sales No. E.96.IX.3.

43. For additional information, see the section titled "Background" in Chapter 1.

44. Todd Hann, Defense Threat Reduction Agency, personal communication, August 13, 2004.

7

Conventional Weapons

Since the Vietnam War significant strides have been made in improving the accuracy of both aircraft navigation systems and systems for the delivery of conventional weapons. Cruise missiles and modern aircraft delivery systems employing high technology such as improved inertial systems, laser guidance, the Global Positioning System (GPS), and television guidance have provided the ability to deliver weapons in an all-weather environment with very high accuracy. Stealth, forward-looking infrared, night-vision goggles, and GPS provided U.S. forces with the ability to operate with impunity at night, along with virtually unchallenged air superiority, in the recent Iraq war. This ability has brought us to an era in which the major limitation for aviation strike operations against hard and deeply buried targets (HDBTs) is target intelligence and the ability to penetrate and destroy HDBTs.

The emerging Global Strike Mission of the Department of Defense requires the capability to deliver rapid, extended-range, precision kinetic (nuclear and conventional) and nonkinetic weapons in support of theater and national objectives.[1] This chapter briefly describes the U.S. family of non-nuclear weapons, current and under development, that support this mission. A summary of these precision-guided munitions is presented in Table 7.1.

- *The Guided Bomb Unit-28 (GBU-28)* is a weapon developed for penetrating hardened, underground Iraqi command centers. The GBU-28 is a 5,000 pound, laser-guided conventional munition that uses a 4,400 pound penetrating warhead (BLU-113) containing 630 pounds of high explosives.
- *The Guided Bomb Unit-24, Advanced Unitary Penetrator (AUP),* is the next-generation, hard-target penetrator munition. It shares the external appearance and flight characteristics of the 2,000 pound BLU-109 but has an advanced, heavy steel penetrator warhead filled with high-energy explosives that can penetrate more than twice as much reinforced concrete as the BLU-109. Performance is enhanced by a void-sensing hard-target smart fuze that detonates the AUP at the optimum point in a target to inflict maximum damage.
- *The BLU 118/B Thermobaric Bomb* added a thermobaric explosive fill for the BLU-109 penetrator. The BLU-118/B is a penetrating warhead filled with an advanced thermobaric explosive that, when detonated, generates higher sustained blast pressures in confined spaces such as tunnels and under-

TABLE 7.1 Summary of Precision-Guided Munitions in U.S. Family of Conventional High-Explosive Weapons, Current and Under Development, That Support the Global Strike Mission

Weapon	Description	
Guided Bomb Unit (GBU)-28/ EGBU-28/ BLU-113	Air-launched, 5,000 pound, laser-guided "bunker buster" with a 4,400 pound penetrating warhead.	
GBU-24B/D/ BLU-116 Advanced Unitary Penetrator (AUP)	Air-launched, 2,000 pound, heavy steel penetrator warhead filled with high-energy explosives and void-sensing hard-target smart fuze that detonates the AUP at the optimum point in a target to inflict maximum damage.	
BLU-118B	Thermobaric bomb, which added a thermobaric explosive fill for the BLU-109 penetrator. The BLU-118/B is a penetrating warhead filled with an advanced thermobaric explosive that, when detonated, generates higher sustained blast pressures in confined spaces such as tunnels and underground facilities. The BLU-118/B uses the same penetrator body as the standard BLU-109 weapon. The significant difference is the replacement of the high-explosive fill with a new thermobaric explosive that provides increased lethality in confined spaces.	
Conventional Air-Launched Cruise Missile (CALCM) Block II Penetrator	Air-launched cruise missile with precision guidance and a 1,200 pound AUP penetrating warhead augmented with two forward shaped charges (BROACH concept) for use against buried and/or hardened targets.	
Joint Standoff Weapon (JSOW)/ Bomb Royal Ordnance Augmenting Charge (BROACH)	Air-launched weapon incorporating the BROACH Multiple Warhead System (MWS), combining an initial penetrator charge (warhead) with a secondary follow-through bomb, supported by multi-event hard-target fuzing. Has a 500 pound class "unitary" warhead providing blast/fragmentation effects as well as enhanced penetration capability against hard targets.	

TABLE 7.1 Continued

Weapon	Description	
Tactical Missile System Penetrator (TACMS-P)	An Advanced Concept Technology Demonstration integration of the Army Tactical Missile System (ATACMS) booster with a Navy reentry vehicle, resulting in an improved capability against buried and/or hardened targets. The TACMS-P range extends to 220 km and will be compatible with the Multiple Launch Rocket System family of launchers.	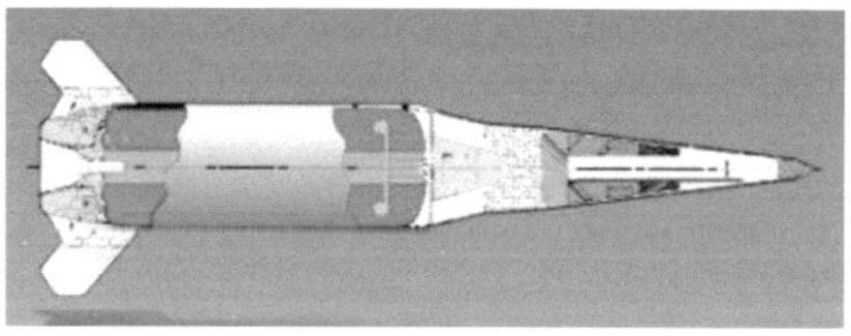
Joint Air-to-Surface Standoff Missile (JASSM)	Conventionally armed, low-observable cruise missile with a 1,000 pound warhead optimized for penetration; it carries a new, high-yield explosive and a hard-target smart fuze.	
Joint Direct Attack Munition (JDAM)	The JDAM will upgrade the existing inventory of Mk-83 1,000 and Mk-84 2,000 pound general-purpose unitary bombs by integrating a guidance kit consisting of an inertial navigation system/Global Positioning System guidance kit. The 1,000 pound variant of JDAM is designated the GBU-31, and the 2,000 pound version is designated the GBU-32. Hard-target penetrators being changed into low-cost JDAMs include the 2,000 pound BLU-109 and the 1,000 pound BLU-110.	
Tactical Tomahawk Penetrator	Tactical Tomahawk missile modified to incorporate a penetrator warhead and the hard-target smart fuze.	

ground facilities. The BLU-118/B uses the same penetrator body as the standard BLU-109 weapon. The significant difference is the replacement of the high-explosive fill with a new thermobaric explosive that provides increased lethality in confined spaces.[2]

• *The Conventional Air-Launched Cruise Missile (CALCM) Block II* is the Precision Strike variant with guidance upgrades and a 1,200 pound AUP warhead augmented with two forward shaped charges for use against buried and/or hardened targets.

• *The Joint Standoff Weapon (JSOW)/Bomb Royal Ordnance Augmenting Charge (BROACH),* a joint Navy/Air Force program, is an air-to-ground weapon designed to attack a variety of targets during day, night, and adverse weather conditions. It includes the integration of the BROACH Multiple Warhead System (MWS) with a 500 pound unitary warhead, providing blast/fragmentation effects as well as enhanced penetration capability against hard targets. The BROACH concept combines two forward shaped charges as part of the penetrator with a secondary follow-through bomb, supported by multi-event hard-target fuzing. The outcome is a warhead and fuze combination that provides more than twice the penetration achievable for equivalent single penetrating warhead types, at an equivalent weight and velocity. The warhead technology can be scaled and configured for a variety of weapon payload and targets requirements.

- *The Tactical Missile System Penetrator (TACMS-P)* is an Advanced Concept Technology Demonstration (ACTD) program to demonstrate integration of the Army Tactical Missile System (ATACMS) booster with a Navy reentry body, resulting in an improved capability against buried and/or hardened targets. The TACMS-P range extends to 220 kilometers and will be compatible with the Multiple Launch Rocket System family of launchers. The reentry body has maneuvering capabilities for increased accuracy.
- *The Joint Air-to-Surface Standoff Missile (JASSM)* is a conventionally armed, low-observable cruise missile. The missile has automatic target recognition, autonomous guidance, precision accuracy, and a 1,000 pound warhead optimized for penetration; it carries a new, high-yield explosive and a hard-target smart fuze.
- *The Joint Direct Attack Munition (JDAM) GBU-31/32,* a guidance tail kit under development to meet both U.S. Air Force and Navy needs, converts existing unguided free-fall bombs into accurate, "smart munitions" for use in adverse weather. With the addition of a new tail section that contains an inertial navigational system and GPS guidance control unit, JDAM improves the accuracy of unguided, general-purpose bombs in any weather condition. The program will produce a weapon with high accuracy and an all-weather, autonomous, conventional bombing capability. JDAM will upgrade the existing inventory of general-purpose and penetrator unitary bombs, and a product improvement may add a terminal seeker to improve accuracy. JDAM can be launched from approximately 25 kilometers from the target, and each is independently targeted.
- *The Tactical Tomahawk Penetrator* is a variation of the BGM-109 Tomahawk, which is an all-weather submarine or ship-launched land-attack cruise missile. After launch, a solid propellant propels the missile until a small turbofan engine takes over for the cruise portion of flight. Tomahawk is a highly survivable weapon. Its detection by radar is difficult because of the missile's small-cross-section, low-altitude flight. Similarly, infrared detection is difficult because the turbofan engine emits little heat. Systems include a GPS receiver; an upgrade of the optical Digital Scene Matching Area Correlation (DSMAC) system; Time of Arrival (TOA) control, and improved 402 turbo engines. A modification of the Navy's tactical Tomahawk missile is being developed that includes a penetrator warhead and the hard-target smart fuze.

NOTES

1. The U.S. Strategic Command is directed to provide capabilities established in the Nuclear Posture Review, full-spectrum global strike, and coordinated space and information operations capabilities to meet national security objectives for both deterrence and decisive action.
2. For additional information, see section entitled "Background" in Chapter 1.

8

Uncertainty in Estimates of Effects

The previous chapters describe earth-penetrator and surface-burst nuclear weapons and summarize the effects of their use. Evaluation of these effects is based on a series of coupled physical and chemical models that describe nuclear earth-penetrator weapons (EPWs), surface-burst weapons, and conventional weapons, the performance of the weapons, the extent of target destruction, related prompt effects and effects of fallout, and the resultant health and environmental effects. For the model calculations, a range of boundary conditions has been assumed. Inevitably, there is uncertainty in such calculations, and the scale of these uncertainties is essential to understanding the results of the calculations. The uncertainties are of three types:

1. *Scenario uncertainty.* This type of uncertainty encompasses the range of parameters calculated for a variety of conditions. In the committee's analysis, the principal factors in scenario uncertainty are the weather at the time of detonation, the distribution of the population, and, for the EPW, heterogeneities of the geologic formations surrounding the target.
2. *Data uncertainty.* For most of the data inputs into the models there is some level of uncertainty. For fundamental physical constants and materials properties, these uncertainties are generally low. Other parameters may be more uncertain, such as the types and volumes of activation products in the fallout. Data uncertainty arises both from random and systematic sources of error, which degrade precision and accuracy, respectively.
3. *Conceptual model uncertainty.* Underpinning all of the calculations are simplified models of the physics, chemistry, and biology of the relevant processes. These include the models of transport and dispersion of the radioactivity, and exposure pathways that lead to a calculated dose and consequent number of fatalities. The conceptual model uncertainty can be large and is the most difficult to quantify, as it is analogous to a systematic source of error in observations.

In any series of calculations, the uncertainty from each of these sources propagates, and grows, through the analysis. To evaluate the sources and magnitude of these uncertainties, the present chapter uses a parametric analysis in which important parameters have been varied across a reasonable range.

The committee has not, however, discussed one of the more important sources of uncertainty in planning the successful defeat of hard and deeply buried targets or their contents (e.g., chemical or biological agents), that is, the need for precise intelligence about the type and configuration of a target.

To assess uncertainties, it is essential to identify the factors to which the model calculations (e.g., estimates of casualties for a given scenario) are more sensitive, as these factors offer the greatest potential contributions to uncertainties. Such factors include wind direction relative to the spatial distribution of a population and the degree to which populations are sheltered. Weather can be accounted for immediately prior to an attack, and sheltering can be accounted for by the timing of an attack (day or night) or perhaps by the issuing of a warning prior to an attack.

For example, based on the Hazard Prediction and Assessment Capability (HPAC) code calculations summarized in Chapter 6, the use of a nuclear EPW instead of an above-surface burst of equivalent military effectiveness (i.e., 25-fold larger yield) is expected to reduce casualties by factors of 2 to 7, 10 to 30, and 15 to 60 for Targets A (urban), B (rural), and C (rural), respectively (the values are for annual estimates of fatalities plus serious injuries). The equivalent military effectiveness coupling factor ranges between 15 and 25. In some of its calculations, the committee compares weapons differing in yield by a factor of 19 because these calculations had already been done and demonstrate the major features of importance. For a given weapon, however, the calculated casualties vary by factors of up to 4 to 8, depending on how well sheltered the population is assumed to be. In addition, Figures 6.9 and 6.11 show that estimates of total casualties can vary by factors as large as 10^1 to 10^2, depending on wind direction.[1]

The general conclusion derived from such comparisons is that the estimated reductions in casualties from a nuclear EPW as opposed to a surface burst of 25-times-higher yield are about 2 to 50 times larger for rural than for urban targets, but the casualty estimates are also more variable by factors of 1 to 2 orders of magnitude in absolute values for the rural than the urban targets. To be sure, all else being equal, use of a weapon with a lower yield is always expected to result in fewer casualties than use of a weapon with a higher yield, for a given set of weather conditions.

SOURCES OF UNCERTAINTY

As discussed in each of the previous chapters, estimates of the effectiveness of as well as the casualties from the attack of a target take into account various sources of uncertainty, which can be summarized as follows.

Positive identification and reliable determination of the three-dimensional coordinates of relevant targets are the main sources of uncertainty considered in Chapter 2 (and again in Chapter 4). Adequate intelligence is required not only to identify those facilities that pose significant threats but also to determine the best modes of attack—which depend on the physical characteristics of the bunker (depth, size, distribution of chambers, hardness, and so on). Uncertainty about these variables affects estimates of target destruction rather than assessment of casualties.

Chapter 3 discusses the heterogeneous nature of low- to medium-strength natural rock formations and the challenge of designing an EPW capable of surviving the lateral and axial forces encountered during impact and penetration of such targets. During the development phase of a given EPW weapon, axial and lateral loading limits of the EPW are determined through extensive component-level shock tests and full-scale EPW system penetration tests. Analytical models are then used to estimate EPW axial and lateral loading for comparison with the EPW survivability limits to assess EPW functionality after the penetration event. The probability of successful penetration is the probability that the EPW will function after it penetrates the target. To determine the probability of successful penetration for a given EPW in a given target, a series of Monte Carlo calculations are run in which each impact parameter and target property variable is given a distribution based on the best information available. The results of the calculations are compared with the loading limitations of the EPW, and a probability of EPW functionality after penetration is determined statistically.

For completeness, it should be noted that in some instances impact parameters and target properties can be severe enough to cause the structural failure of the EPW's casing. Safety-related tests conducted on various EPW development projects where EPWs have been tested beyond loading limits indicate that it is extremely unlikely that a second-order detonation of the high explosives will occur, even if the EPW case is ruptured at impact or during penetration. It is possible, of course, that a small number of pieces of nuclear material could be dispersed in the immediate area if the EPW case ruptured on the surface.

Given the impact parameters of the EPW system, the range of forces to which the EPW could be subjected are calculated for a wide range of target properties and impact conditions in order to assess the utility of the weapon. To estimate the uncertainty in EPW functionality, one can then evaluate the range of forces to which an EPW would be subjected, for a wide range of target properties and impact conditions, associated with the known (or estimated) heterogeneity of a given target.

As extrapolated from research done in support of previous design work (e.g., for the Pershing-II EPW, the W61, and the strategic earth-penetrator weapon (SEPW) described in Chapter 3), estimates of the B61-11's functionality are high for attacks on its design targets, but very low for its attacks on a variety of other complex and harder targets. The robust nuclear earth penetrator (RNEP) weapon concepts currently under consideration could conceivably achieve a probability of functionality of perhaps 40 to 70 percent relative to axial and lateral loads, while an advanced penetrator design could achieve a probability of functionality greater than 90-plus percent.[2] The committee emphasizes that these are little more than initial judgments, however. Although informed by a sizeable amount of experimental data, they are supported by only a limited amount of fundamental analysis. The dearth of realistic assessments of subsurface heterogeneity is a notable gap in current analysis.

Chapters 4 and 6 list several sources of uncertainty influencing the calculated effects of nuclear weapons. The source term depends on accurately knowing weapon yield and height or depth of burst, as well as the target environment and consequent effects of secondary (neutron) activation. The reliable prediction of transport, both for nuclear fallout and for released chemical or biological agents, depends on accurate forecasts of wind direction, wind speed, and rain, including the effects of terrain (natural topography and, for urban environments, buildings). Health and environmental consequences are currently evaluated only for military casualties, not for the population at large (e.g., including the young, old, and infirm), and even then are considered reliable to only a factor of two. Also, the population databases are necessarily static, and, even if reliable in an average sense, do not reflect actual locations and movements of individuals. Chapter 5 does note that the integrated dose contours reproduce to within a factor of two the nuclear test results from which the empirical models were derived; however, it is important to acknowledge that the nuclear tests were conducted in good weather and under relatively stable atmospheric conditions.[3]

Additional sources of uncertainty listed in Chapter 4 include the hardness of the target, the effects of the accuracy of weapons delivery (circular error probable, CEP), and details of fallout assessment (e.g., the relationship between amount of radioactivity and size for particles in the fallout). Chapter 6 documents the importance of target location (urban versus rural, as noted above) and also discusses uncertainties in estimates of casualties resulting from the attack of facilities containing chemical or biological agents. It is noteworthy that a nuclear weapon offers an advantage for defeating biological or chemical weapons agents within the facility only if the warhead can be detonated within the chamber containing these agents; otherwise, the special effects by which the nuclear detonation can neutralize the agents are likely lost.

In summary, estimates of casualties are based on combined modeling of the source, transport, and health consequences of nuclear weapons effects. Uncertainties in the casualty estimates are difficult to evaluate because the underlying processes are poorly understood. In particular, the fidelity with which the transport of aerosols (mimicked by way of advection diffusion) is modeled remains unclear, and the prediction of health effects—based on empirical measurements, but limited by the relative lack of data and unsupported by chemistry- and biology-based theory—is modeled with a reliability that is hard to assess.

MEASURES OF UNCERTAINTY: SENSITIVITY, PRECISION, AND ACCURACY

A part of the overall uncertainty regarding estimates of the effects of weapon use can be inferred from the calculated sensitivity of the results to varying conditions and assumptions. As noted in Chapter 6, expected casualties can be expected to vary by up to a factor of 10 for urban targets, and by a factor of nearly 100 for rural targets, depending on assumptions that are difficult to validate a priori. In terms of absolute numbers of fatalities and injuries, however, attacks on urban targets dominate overwhelmingly. That is, estimated casualties from the use of a nuclear weapon in an urban area are typically in the 10^5 to 10^6 range, whereas casualties from the rural targets considered in this study can be as low as in the 10^1 to 10^2 range (albeit potentially extending to the 10^5 range, depending on circumstances and assumptions). For this reason, it is important to consider absolute as well as relative variations in evaluating calculated fatalities and severe injuries.

In addition to sensitivity, both the precision and the accuracy of the modeling are considered. To determine precision, the committee evaluated reproducibility by comparing results using primarily two simulation tools, the Defense Threat Reduction Agency's (DTRA's) HPAC code and Lawrence Livermore National Laboratory's (LLNL's) K-Division Defense Nuclear Agency Fallout Code (KDOFC). As discussed in Chapters 5 and 6, the numbers of casualties obtained with the two codes generally agree to within 10 to 30 percent for a wide range of scenarios. Given that variations of orders of magnitude in estimated casualties are being considered, the committee considers the codes to yield reproducible and therefore mutually consistent results. This level of agreement no doubt reflects the fact that the codes have been calibrated against the same set of available field measurements (primarily from 10 aboveground nuclear tests).

Absolute accuracy is much harder to quantify, and so the study breaks the problem down into estimating accuracy for key components of the model calculation. For example, the sensitivity of calculated casualties to wind direction, as described in Chapter 6, can be convolved with uncertainties in the forecast winds for a particular location and time in order to deduce one component of the ultimate accuracy of the final results. A full evaluation of meteorological uncertainty is intrinsically difficult because length scales for fluctuations in wind velocity are in the 10^1 to 10^2 meter range (e.g., "outer scale" of atmospheric turbulence),[4] whereas the smallest grid spacing considered in the models is 10^3 meters, and this typically represents interpolations across even greater distances.

How good are the predicted wind directions from which to estimate casualties using such tools as HPAC or KDFOC? A qualitative answer is given by recent reports that highlight the role of wind-vector fields in determining the uncertainties in modeling atmospheric plumes.[5] Anecdotally, it is also known that the primary wind direction changed by 180° within 18 hours of the Chernobyl incident, thus spreading nuclear contaminants in more directions than might have otherwise been expected, and that the smoke plume from the World Trade Center site was directed opposite to the normal wind direction shortly after the September 11, 2001, attack.[6]

A more quantitative answer comes from comparing the results of modern numerical weather predictions against field observations. Experience in the Pacific Northwest, for example, shows systematic biases (mean errors) of ±10 to 20 degrees and mean absolute-value errors of ±40 to 50 degrees for near-surface (10 meters) winds at 12 to 24 hours into forecasts, depending on spatial resolution and forecast hour (the near-surface winds are expected to be less subject to directional shear under unstable or neutral conditions than winds at higher elevation).[7,8] Similarly, mean absolute-value errors of ±30 to 40 degrees are documented for a model applied to the Savannah River Site in South Carolina, though with excursions reaching 60 to 70 degrees at 12 to 24 hours into the forecast.[9] Not surprisingly, the statistics can be somewhat worse for topographically complex areas, such as the region around Salt Lake City, Utah, although mean absolute errors of ±40 to 50 degrees remain possible in many specific locations when extensive modeling and updating are pursued.[10,11] In general, wind directions are more poorly forecast at lower than at higher wind speeds, and at lower (near-surface) than higher levels of the atmosphere.

Such discrepancies between forecast and observed wind-vector fields (wind directions, in particular) are widely recognized in the community that models atmospheric dispersion and photochemistry.[12] For instance, one document mentions average root-mean-square errors of ±36 to 57 degrees in wind direction (largest values near Earth's surface, smallest values at pressure levels below 100 millibars), but with directional excursions in estimated surface-wind mean errors (bias) as large as 106 degrees (the analysis pertains to the March 10, 1991, meteorological event of Khamisiyah, Iraq). Similar modeling applied to northern Europe reveals systematic biases (mean errors) and root-mean-square errors of up to ±13 to 15 degrees and ±35 to 41 degrees, respectively, at the lowest levels and beyond 24 hours of forecast.[13]

Studies of ensemble forecasts do not alter the substance of these findings, but instead reinforce the conclusion that reliable forecasts of dispersion are a challenge: Straume reports root-mean-square deviations in wind direction between 15 and 48 degrees, for example.[14] This is not surprising, given the errors to which models are subject and the difficulty—and therefore diminished reliability—with which wind-field directions are measured.[15]

Overall, it appears that systematic biases (mean errors) of ±10 to 20 degrees and root-mean-square (or mean absolute-value) errors of 30 to 50 degrees should be expected for wind direction forecasts relevant to the present study. Referring to Figures 6.9 (a) and (b), an uncertainty of ±20 degrees encompasses the difference between maximum and minimum estimates of total casualties for Target A (urban), and corresponds to more than a 20-fold difference in casualties for Target B (rural). In both cases, these uncertainties amount to differences on the order of 10^6 in calculated values of fatalities and serious injuries. In light of the present results, Figures 6.11 (a) and (b) make the point that although a low-yield nuclear EPW is expected to cause fewer fatalities than a higher-yield above-surface nuclear burst, the differences in forecasted deaths for surface versus EPW detonations are comparable to the variability associated with a ±15 to 40 degree uncertainty in forecast wind directions. The general conclusion is that uncertainties in wind direction can result in casualty estimates varying by one order of magnitude or more.

For comparison, the degree to which the population is—and remains—sheltered, or not, affects the results by a smaller amount. Other factors being equal, sheltering is assumed always to lead to a reduction of casualties because of the diminished influence of fallout.[16] HPAC analyses show that the ratio of calculated unsheltered/sheltered casualties (fatalities plus serious injuries) varies between 2 and 8 for virtually all cases of a nuclear EPW as considered here. Again, the smallest *ratios* apply to the urban target (A) but correspond to the largest differences in *actual values* (a reduction in casualties of between 100,000 and 600,000 if the population stays indoors rather than being outside). The magnitude of these differences in casualty reduction numbers is because of the fact that the absolute number of expected casualties is always high, ranging from 200,000 to 2 million for the urban nuclear EPW scenarios evaluated for this study. The largest ratios apply to use of a low-yield weapon on a rural target (B or C) but involve relatively few casualties in the first place (e.g., a reduction from 200 to 30 and from 20 to 6 calculated fatalities plus serious injuries for two of the scenarios). *Absolute values* of fatalities and serious injuries are systematically higher for the above-surface nuclear explosion attacks considered in the study, but the *ratios* of calculated unsheltered/sheltered casualties are in the same range as for the nuclear EPW attacks of equivalent military effectiveness (i.e., yield reduced by a factor of 25).

As it seems a priori impossible to predict convincingly the degree to which a population will, or will be able to, stay indoors after a nuclear attack, the casualty estimates from such tools as HPAC and KDFOC must be considered globally unreliable to a factor of 2 to 8 as a result of this uncertainty alone. Additional sources of uncertainty were not evaluated in as great detail, but based on the discussions in Chapters 2 to 6, these may again amount to an overall factor of about 2 to 5 (e.g., estimates of the health effects of a given fallout dose for the actual population, rather than for the static, military-age population assumed in the calculations, are by themselves likely to be unreliable by a factor of 2).

Thus, potential errors aside from those associated with weather forecasts can reach an order of magnitude in calculated casualties. Given that uncertainties in wind direction additionally cause variations of one order of magnitude or more, the aggregate uncertainty for calculations of casualties caused by the range of nuclear attack scenarios considered here must be squarely placed in the range of 10^1 to 10^2 (factors of 10 to 100), with the lower range applying to urban targets for which casualties will with little doubt be very high.

SUMMARY

Current analytical tools have an overall propagated uncertainty no smaller than one order of magnitude (factor of 10), and likely in the range of 10^1 to 10^2, for estimates of casualties resulting from a nuclear attack. This conclusion is based both on evaluation of the underlying calculations (source terms, transport models, grid resolution, and so on) and their experimental validation, and on a review of the variability in results that can be obtained for different scenarios when considering plausible ranges in parameters.

At least three key sensitivities affect estimates of military effectiveness and casualties associated with the use of a nuclear EPW or a nuclear burst weapon:

1. Target location, especially urban versus rural, as illustrated above;
2. Accuracy of weapons delivery (CEP) and precise knowledge of target location and structure, as military effectiveness depends closely on a combination of accurate delivery and yield; and
3. Estimates of the source, transport, and influence on populations of the effects of a nuclear explosion, as these can be highly variable (by factors of up to ~10^1 to 10^3, depending on assumptions).

One additional sensitivity affects the nuclear EPW:

4. EPW functionality after penetration, especially as influenced by target heterogeneity and the associated uncertainty (e.g., local geology, or complex structures in urban areas).

Some conclusions involve relatively little uncertainty. All other factors being equal, the use of a lower-yield weapon causes fewer casualties than use of a higher-yield weapon, for example, and a nuclear attack on an urban target must be expected to result in large numbers of fatalities and serious injuries (hundreds of thousands to millions, for the scenarios considered in this study). Relative variations in calculated casualties are therefore smaller for urban than rural targets, though the absolute numbers of predicted deaths can differ by large amounts (hundreds of thousands) for urban areas depending on factors that are not readily controlled (e.g., degree of sheltering).

For urban and rural targets, respectively, the use of a nuclear EPW is calculated to reduce casualties by a factor of ~2 to 7 and 10 to 60 relative to an aboveground nuclear burst with a yield increased by a factor of 25. This calculated 10^1 to 10^2 reduction in fatalities and serious injuries is comparable to the effect of the aggregate uncertainties that the committee has derived for the modeling tools.

NOTES

1. The use of exponential notation ($10^0 = 1$, $10^1 = 10$, $10^2 = 100$, $10^3 = 1{,}000$, and so on) implies that the values indicate only orders of magnitude (powers of ten) and are typically uncertain by factors of at least 2 to 5.

2. Details are provided in the following classified reports: *W86 Warhead Status Report (U)*, SAND 83-1642, RS 3151/83/033, Sandia National Laboratories, Albuquerque, N.Mex. (11/1/1983); *Strategic Earth Penetrator Joint DOD/DOE Phase 2 Study (U)*, Robert Blankert, Air Force Material Command, RS 2907/01/00268, NWIC-TR-94-2 (9/1/1998); and *W61 Weapon Development Report (U)*, SAND 91-2243, RS 3151/91/00024, Sandia National Laboratories, Albuquerque, N.Mex. (3/1/1992).

3. Ted F. Harvey, Lawrence Livermore National Laboratory, March 23, 2004, personal communication.

4. K.A. Hart, W.J. Steenburgh, D.J. Onton, and A.J. Siffert. 2004. "An Evaluation of Mesoscale-Model-Based Model Output Statistics (MOS) During the 2002 Olympic and Paralympic Winter Games," *Weather and Forecasting*, Vol. 19, pp. 200-218.

5. National Research Council, 2003a, *To Live on an Active Earth: Perspectives on Earthquake Science*, National Academies Press, Washington, D.C.; National Research Council, 2003b, *Tracking and Predicting the Atmospheric Dispersion of Hazardous Material Releases*, National Academies Press, Washington, D.C.

6. Fred A. Mettler, Jr., New Mexico Federal Regional Medical Center, March 23, 2004, personal communication.

7. To avoid ambiguities in comparing directions greater or less than 360 degrees, the errors are typically reported out of a total possible range of 180 degrees and are therefore listed here as ± values out of 360 degrees.

8. E.P. Grimit and C.F. Mass, 2002, "Initial Results of a Mesoscale Short-Range Ensemble Forecasting System over the Pacific Northwest," *Weather and Forecasting*, Vol. 17, pp. 192-205; C.F. Mass, D. Ovens, K. Westrick, and B.A. Cole, 2002, "Does Increasing Horizontal Resolution Produce More Skillful Forecasts? The Results of Two Years of Real-Time Numerical Weather Prediction over the Pacific Northwest," *Bull. Am. Meteorol. Soc.*, Vol. 83, pp. 407-430.

9. R.L. Buckley, A.H. Weber, and J.H. Weber. 2004. "Statistical Comparison of Regional Atmospheric Modelling System Forecasts with Observations," *Meteorol. Appl.*, Vol. 11, pp. 67-82.

10. Given the level of interest associated with the Olympic Games, from media, commercial enterprises, and participants, among others, Hart et al.'s 2004 study could be considered to describe a best-case scenario for current modeling capabilities applied to a complex terrain.

11. K.A. Hart, W.J. Steenburgh, D.J. Onton, and A.J. Siffert. 2004. "An Evaluation of Mesoscale-Model-Based Model Output Statistics (MOS) During the 2002 Olympic and Paralympic Winter Games," *Weather and Forecasting*, Vol. 19, pp. 200-218.

12. A. Russell and R. Dennis, 2000, "NARSTO Critical Review of Photochemical Models and Modeling," *Atmos. Environ.*, Vol. 34, pp. 2283-2324; S. Zhong and J. Fast, 2003, "An Evaluation of the MM5, RAMS, and Meso-Eta Models at Sub-Kilometer Resolution Using VTMX Field Campaign Data in the Salt Lake Valley," *Monthly Weather Review*, Vol. 131, pp. 1301-1322; D.P. Bacon, N.N. Ahmad, Z. Boybei, T.J. Dunn, M.S. Hall, P.C.S. Lee, R.A. Sarma, M.D. Turner, K.T. Waight III, S.H. Young, and J.W. Zack, 2000, "A Dynamically Adapting Weather and Dispersion Model: The Operational Multiscale Environment Model with Grid Adaptivity (OMEGA)," *Monthly Weather Review*, Vol. 128, pp. 2044-2076.

13. Z. Boybei, N.N. Ahmad, D.P. Bacon, T.J. Dunn, M.S. Hall, P.C.S. Lee, R.A. Sarma, and T.R. Wait. 2001. "Evaluation of the Operational Multiscale Environment Model with Grid Adaptivity Against the European Tracer Experiment," *J. Appl. Meteorol.*, Vol. 40, pp. 1541-1558.

14. A. Straume. 2004. "A More Extensive Investigation of the Use of Ensemble Forecasts for Dispersion Model Evaluation," *J. Appl. Meteorol.*, Vol. 40, pp. 425-445.

15. M.H. Freilich and R.H. Dunbar, 1999, "The Accuracy of the NSCAT 1 Vector Winds: Comparisons with National Data Buoy Center Buoys," *J. Geophys. Res.*, Vol. C5, pp. 11231-11246; D. Orell, L. Smith, J. Barkmeijer, and T.N. Palmer, 2001, "Model Error in Weather Forecasting," *Nonlinear Processes in Geophysics*, Vol. 8, pp. 357-371.

16. In contrast, casualties from earthquakes are often observed to be greatly magnified due to the effects of building collapses when a large fraction of a population is indoors—for example at night.

9

Conclusions

The previous chapters present the results of the work carried out to address the committee's charge, including a literature review, calculations, and analysis of information presented by many experts. The process led the committee to reach several conclusions. Listed below are the nine conclusions that the committee believes are most important to addressing the issues raised in the charge, followed by additional conclusions grouped by general topic.

MOST IMPORTANT CONCLUSIONS

Conclusion 1. Many of the more important strategic hard and deeply buried targets (HDBTs) are beyond the reach of conventional explosive penetrating weapons and can be held at risk of destruction only with nuclear weapons. Many—but not all—known and/or identified hard and deeply buried targets can be held at risk of destruction by one or a few nuclear weapons.

Conclusion 2. Nuclear earth-penetrator weapons (EPWs) with a depth of penetration of 3 meters capture most of the advantage associated with the coupling of ground shock. While additional depth of penetration increases ground-shock coupling, it also increases the uncertainty of EPW survival. To hold at risk hard and deeply buried targets, the nuclear yield must be increased with increasing depth of the target. The calculated limit for holding hard and deeply buried targets at risk of destruction with high probability using a nuclear EPW is approximately 200 meters for a 300 kiloton weapon and 300 meters for a 1 megaton weapon.

Conclusion 3. Current experience and empirical predictions indicate that earth-penetrator weapons *cannot* penetrate to depths required for total containment of the effects of a nuclear explosion.

Conclusion 4. For the same yield and weather conditions, the number of casualties from an earth-penetrator weapon detonated at a few meters depth is, for all practical purposes, equal to that from a surface burst of the same weapon yield. Any reduction in casualties due to the use of an EPW is attributable primarily to the reduction in yield made possible by the greater ground shock produced by buried bursts.

Conclusion 5. The yield required of a nuclear weapon to destroy a hard and deeply buried target is reduced by a factor of 15 to 25 by enhanced ground-shock coupling if the weapon is detonated a few meters below the surface.

Conclusion 6. For attacks near or in densely populated urban areas using nuclear earth-penetrator weapons on hard and deeply buried targets (HDBTs), the number of casualties can range from thousands to more than a million, depending primarily on weapon yield. For attacks on HDBTs in remote, lightly populated areas, casualties can range from as few as hundreds at low weapon yields to hundreds of thousands at high yields and with unfavorable winds.

Conclusion 7. For urban targets, civilian casualties from a nuclear earth-penetrator weapon are reduced by a factor of 2 to 10 compared with those from a surface burst having 25 times the yield.

Conclusion 8. In an attack on a chemical or biological weapons facility, the explosive power of conventional weapons is not likely to be effective in destroying the agent. However, the BLU-118B thermobaric bomb, if detonated within the chamber, may be able to destroy the agent. An attack by a nuclear weapon would be effective in destroying the agent only if detonated in the chamber where agents are stored.

Conclusion 9. In an attack with a nuclear weapon on a chemical weapons facility, civilian deaths from the effects of the nuclear weapon itself are likely to be much greater than civilian deaths from dispersal of the chemical agents. In contrast, if the target is a biological weapons facility, release of as little as 0.1 kilogram of anthrax spores will result in a calculated number of fatalities that is comparable on average to the number calculated for a 3 kiloton nuclear earth-penetrator weapon.

OTHER CONCLUSIONS

Hard and Deeply Buried Targets and Nuclear Earth-Penetrator Weapons

In potential adversary nations, there is a large (2,000) and growing number of identified, strategically important facilities that are sheltered in underground bunkers.

Weapon Effects

Because of the limitations on the penetration depth of penetrating weapons in the stockpile today, as well as those of the robust nuclear earth penetrator (RNEP) weapon currently under study, effects of their use would not be contained.[1] Depending on many variables, including weapon yield, proximity of the target to urban areas, distribution of the civilian population and warning time available for sheltering or evacuation, ambient wind profile, and other weather conditions, this study's calculations of numbers of deaths and serious injuries resulting from attacks on representative targets near or in urban areas range from less than 10^3 to greater than 10^6.

Nuclear EPWs (300 kilotons to 1 megaton) can hold at risk HDBTs of interest (1 kilobar hard) at up to ~100 meters to 300 meters depth of burst in granite with high probability of damage (PD greater than 0.95) if delivered with high precision.

Targets buried up to 85 meters can be held at risk (PD greater than 0.95) by precision low-yield nuclear weapons (less than 10 kilotons).

A ~100 kiloton weapon with moderate accuracy (~100 meter circular error probable (CEP)) detonated at its fallout-free height of burst (HOB)—that is, with no local fallout in the absence of rain—can be highly effective (PD greater than 0.95) against surface or near-surface, moderately hard (~500 pounds per square inch (psi) overpressure) point targets. To be highly effective against targets that are harder than about 2000 psi, detonation must occur lower than the fallout-free HOB, regardless of the yield or accuracy of the weapon. For a surface-burst weapon, the yield required to destroy with high probability (PD greater than 0.95) very hard surface or near-surface point targets (e.g., missile silos requiring 1,000 psi to 5,000 psi overpressure) is highly dependent on accuracy (e.g., 250 kilotons requires a CEP of few tens of meters; 1 megaton requires a CEP of ~100 meters). The importance of the accuracy of weapon delivery (CEP) increases with increasing target depth of burial up to about 150 meters. Beyond this depth, the importance of accuracy diminishes relative to that of increased yield.

Differences in assumptions regarding sheltering and evacuation of the population can alter the estimated number of casualties by a factor of 2 to 8. Wind patterns can have an enormous effect on the number of casualties resulting from fallout. For targets in large urban centers, fatalities from acute and latent effects from fallout can vary by more than a factor of 10, and for targets outside cities, fatalities from exposure to fallout can vary by more than a factor of 100, depending on population distribution and which way the wind blows. National leaders can attempt to minimize casualties by choosing time of day and timing attacks for favorable forecasted wind patterns, but the predictability of weather is limited and there may be constraints on the ability to wait until forecasts are favorable.

Design

Credible empirical equations are available for estimating depth of penetration and resulting EPW axial deceleration. Maximum depths of penetration are estimated to be between 7 meters for medium-strength rock to 70 meters for silty clay.

The greatest uncertainty regarding EPW survival concerns the heterogeneous nature of target geologies. Rugged EPW designs will enhance EPW survivability. A depth of penetration of about 3 meters achieves most of the benefits of effective energy coupling, and limiting detonation to that depth avoids the uncertainties associated with geologies below that depth.

Collateral damage from a nuclear earth-penetrator weapon cannot be avoided entirely, but it could potentially be reduced by new design concepts combining deeper-penetration, lower-yield, and low-fission-fraction nuclear design for reduced radioactivity. To achieve such reductions, innovative concepts must be developed for achieving combinations of penetration depth and yield combinations that would substantially diminish the radioactive fallout from a nuclear earth-penetrator weapon attack on an HDBT. Use of such a weapon with reduced yield requires more precise and reliable intelligence and greater delivery accuracy than for an above-surface nuclear burst of comparable military effectiveness.

Chemical and Biological Facilities

To destroy chemical or biological agents in an attack, a weapon must detonate essentially within the chamber in which the agent is stored. In general it must detonate in flight, having penetrated through the protective cover of the underground storage. Similarly, if the agents are not in a single room, but are in adjacent tunnels, no more than a region of a single tunnel could be irradiated by a single nuclear explosion.

If chemical or biological agents are released in an EPW attack, there are several ways to reduce casualties and fatalities. Protective clothing and masks can protect people in the high-hazard areas. Those regions can also be treated chemically to decontaminate them. Chemical agents can be rendered inert by exposure to sunlight, heat, or rain, and neutralizing injections exist in some cases. All exposed biological agents are eventually destroyed by ultraviolet exposure, and the effects of many are preventable by vaccines or are treatable by other medical countermeasures. However, the level of "cleanliness" required by military and public health officials for occupancy of a contaminated area is still under substantial discussion and debate.

Given the same target, using conventional rather than nuclear weapons to destroy a chemical storage facility (surface or buried) most likely will cause fewer casualties in either a populated or an unpopulated area, even if there is a potential for release of the chemicals. The same is not necessarily true for a biological storage facility.

Codes

The codes and models used to estimate environmental and health effects were designed for Cold War scenarios, have many limitations, and are often ill-suited for today's national security environment. In particular, the mind-set in designing tools was often based on particular warfighting modes, involving massive nuclear exchanges or tactical encounters. Modern tools need to emphasize the effects of single releases of weapons of mass destruction in a variety of urban or rural environments characterized by detailed meteorology and terrain.

Documentation of the widely used codes and models is sparse, and so users often do not understand the assumptions underlying the calculations. In particular, the range of applicability and of uncertainty is often left to the user's imagination.

Uncertainty

In the calculations for this study, the probability of finding, identifying, and characterizing the target; weapons system survival and arrival at the target; and weapon penetration and detonation are assumed to be 1.0. These assumptions are recognized as unrealistic. In addition, many cautions are needed regarding conclusions based on the model runs.

The model runs show the following:

1. For urban targets, the use of a nuclear EPW is calculated to reduce casualties by a factor of ~2 to 10 relative to an aboveground nuclear burst whose yield is 25 times larger than that of the EPW. Estimates of this factor can vary by up to 4 to 8, depending on assumptions.

2. For rural targets, the use of a nuclear EPW is estimated to reduce casualties by a factor of 10 to 100 relative to an aboveground nuclear burst, with the variability in the modeling results extending into the 10^2 range.

3. At least four key sensitivities affect estimates of military effectiveness and casualties associated with use of a nuclear EPW:

- Target location, especially urban versus rural, as illustrated above;
- Accuracy of weapons delivery (CEP) and precise knowledge of target location and structure, as military effectiveness depends closely on a combination of CEP and yield;

- Functionality on penetration, especially as influenced by target heterogeneity and its uncertainty (e.g., local geology, or complex structures in urban areas); and
- Estimates of the source, transport, and influence on populations of the effects of a nuclear explosion; the estimates are highly variable (by factors of up to ~10^1 to 10^3, depending on assumptions).

Current analytical tools have an overall propagated uncertainty of about one order of magnitude (factor of 10) in the estimated casualties. This conclusion is based both on evaluation of the underlying calculations (source terms, transport models, grid resolution, and so on) and their experimental validation and on a review of the variability in results that can be obtained for different scenarios when considering plausible ranges in parameters.

NOTE

1. The presence of an unstemmed penetration hole ensures massive venting.

Appendixes

A

Committee and Staff

John F. Ahearne ***(Chair)*** is the director of the Ethics Program at the Sigma Xi Center for Sigma Xi, The Scientific Research Society; a lecturer in public policy at Duke University; and an adjunct scholar at Resources for the Future. His professional interests are reactor safety, energy issues, resource allocation, and public policy management. He has served as commissioner and chair of the U.S. Nuclear Regulatory Commission, system analyst for the White House Energy Office, Deputy Assistant Secretary for Energy, and Principal Deputy Assistant Secretary for Defense. Dr. Ahearne currently serves on the Department of Energy's Nuclear Energy Research Advisory Committee. In addition, he has been active in several National Research Council (NRC) committees examining issues in risk assessment. He is a fellow of the American Physical Society, American Association for the Advancement of Science, American Academy of Arts and Sciences, and Society for Risk Analysis, and a member of Sigma Xi, the American Nuclear Society, and the National Academy of Engineering. He received his B.E.P. and M.S. degrees from Cornell University and a Ph.D. in physics from Princeton University.

Lynn R. Anspaugh is a research professor in radiobiology at the University of Utah, where his research interests range from trace elements in human metabolism to reconstruction of radiation doses from early fallout of nuclear weapons tests. Dr. Anspaugh began his career as a National Science Foundation graduate fellow in radiological physics at the University of California, Berkeley, after which he went on to work as a biophysicist at Lawrence Livermore National Laboratory (LLNL) from 1963 to 1996. He served in various capacities while at LLNL; his last position was as director of the Dose Reconstruction Program. In addition to being a member of the American Association for the Advancement of Science, Sigma Xi, and many other professional organizations, Dr. Anspaugh is a fellow of the Health Physics Society. He obtained his Ph.D. degree in biophysics from the University of California, Berkeley.

Rodney C. Ewing is the Donald R. Peacor Collegiate Professor in the Department of Geological Sciences at the University of Michigan, where he heads a research program on radiation effects and nuclear waste management. He also holds faculty appointments in materials science and engineering and in nuclear engineering and radiological sciences. He is an Emeritus Regents' Professor at the

University of New Mexico in the Department of Earth and Planetary Sciences, where he was a member of the faculty from 1974 to 1997 and chair of the department from 1979 to 1984. Dr. Ewing is a fellow of the Geological Society of America and the Mineralogical Society of America and past president of the International Union of Materials Research Societies and Mineralogical Society of America. He is also an adjunct professor at the University of Aarhus in Denmark. Dr. Ewing's research interests center on mineralogy and materials science and include the long-term durability of radioactive waste forms. He has served on numerous scientific boards and advisory committees, including the NRC Board on Radioactive Waste Management (since 2001). Dr. Ewing received a B.S. in geology from Texas Christian University and an M.S. and a Ph.D. in mineralogy from Stanford University.

Steven A. Fetter is a professor in the School of Public Policy at the University of Maryland, College Park, where he has taught courses in national security policy, environmental policy, and quantitative analysis since joining the school's faculty in 1988. He has been a visiting fellow of the Center for International Security and Cooperation at Stanford University, the Center for Science and International Affairs at Harvard University, the Plasma Fusion Center at the Massachusetts Institute of Technology, and Lawrence Livermore National Laboratory. He has also served as a special assistant to the Assistant Secretary of Defense for International Security Policy and was a Council on Foreign Relations fellow in the State Department. Dr. Fetter's research interests span a wide range of fields, including nuclear weapons, arms control, and nonproliferation policy; nuclear power and the health effects of radiation; and climate change and energy supply. He has served on numerous scientific boards and advisory committees, including the NRC Committee on International Security and Arms Control (since 1995) and the Department of Energy's Nuclear Energy Research Advisory Committee. Dr. Fetter received an S.B. in physics from the Massachusetts Institute of Technology and a Ph.D. in energy and resources from the University of California, Berkeley.

Richard L. Garwin is an emeritus fellow at the IBM Thomas J. Watson Research Center. From 1994 until March 2004, he was the Phillip D. Reed Senior Fellow for Science and Technology at the Council on Foreign Relations, New York. A member of the National Academy of Sciences, National Academy of Engineering, and Institute of Medicine, his expertise in experimental and computational physics includes contributions to nuclear weapons design, instruments and electronics for nuclear and low-temperature physics, computer elements and systems, superconducting devices, communications systems, behavior of solid helium, and detection of gravitational radiation. Dr. Garwin was a member of the President's Science Advisory Committee from 1962 to 1965 and 1969 to 1972 and a member of the Defense Science Board from 1966 to 1969. He currently is an active member of the JASONs and consults for Los Alamos National Laboratory and Sandia National Laboratories. In 1998, he was a member of the nine-person Rumsfeld Commission—the Commission to Assess the Ballistic Missile Threat to the United States. He has written extensively on nuclear-weapons-related issues over the course of several decades, particularly on the question of maintaining the nuclear stockpile under a comprehensive test ban regime. From 1994 until August 2001, he chaired the State Department's Arms Control and Nonproliferation Advisory Board. He is a fellow of the American Physical Society and the American Academy of Arts and Sciences and a member of the American Philosophical Society. He received a B.S. and a D.Sc. from Case Western University and an M.S. and a Ph.D. in physics from the University of Chicago.

Sydell P. Gold is senior vice president at Science Applications International Corporation (SAIC), where she is responsible for SAIC's Defense Threat Reduction Agency's (DTRA's) business activities as DTRA account manager, and for developing new business opportunities for SAIC. Previously, she was also the deputy sector manager, assisting the management of the Advanced Technology and Analysis Sector, a more than $350 million organization with over 2,000 employees, specializing in systems design and engineering and computational and laboratory analysis and research. Prior to joining SAIC in 1992, Dr. Gold served for 10 years with the Office of the Assistant Secretary of the Air Force (Acquisition) as Deputy Assistant Secretary (Staff Support and Analysis) (acting) and as Deputy to the Assistant Secretary. Previously, she had served as a member of the professional staff at the National Security Council, as a technical staff member at Lawrence Livermore National Laboratory performing analyses of nuclear weapon and related security issues, and at Sandia National Laboratories utilizing applied mathematics and systems analyses for national security and nondefense issues. Dr. Gold received a B.A. from Barnard College of Columbia University, an M.S. from the University of New Mexico, and a Ph.D. in mathematics from the University of California, Berkeley.

Eugene G. Grewis is an independent consultant in areas pertaining to nuclear weapons and related defense activities. He retired from the U.S. civil service in 1994 after 33 years, during which his experience included technical, managerial, and operational efforts with the entire nuclear stockpile as well as selected efforts in chemical and biological defense and other special programs. His career spanned 30 years with the Department of Defense and 3 years with the Department of Energy, progressing from hands-on engineering to senior-level management of weapons programs, interdepartmental liaison, nuclear materials, and defense capabilities in the areas of threat definition, evaluation of systems alternatives, resource planning, safety, system operations, and logistics support. Mr. Grewis has received the Department of Energy Exceptional Service Medal, the Navy Superior Civilian Service Medal, and the Navy Meritorious Civilian Service Medal. He received a B.S. in electrical engineering from Washington University at St. Louis.

Theodore M. Hardebeck recently joined Science Applications International Corporation as vice president and director of science, technology, and strategy. He previously served as associate director, concepts and assessments, and as the Commander's science and technology advisor at the U.S. Strategic Command (USSTRATCOM). Dr. Hardebeck's background is in nuclear weapons issues relating to network-centric military planning and analysis. At USSTRATCOM, he led a comprehensive examination of issues involving guidance, target base, weapon requirements, and stability, the results of which provided much of the foundation of the 1991 Presidential Nuclear Initiative. Dr. Hardebeck received a B.S. in mathematics and physics from Ball State University and an M.S. and a Ph.D. in mathematics from Case Western Reserve University.

Raymond Jeanloz is a professor of Earth and planetary science and of astronomy at the University of California, Berkeley, and a recently elected member of the National Academy of Sciences. His research interests include the properties of materials at high pressures and temperatures and the nature of planetary interiors. Dr. Jeanloz has served on numerous scientific boards and advisory committees, including as chair of the NRC Board on Earth Sciences and Resources (from 2000 to 2002) and as a member of the NRC Committee on International Security and Arms Control (since 2002). He is a fellow of the American Academy of Arts and Sciences, the American Association for the Advancement of Science, and the American Geophysical Union. Dr. Jeanloz received a B.A. in geology from Amherst College and a Ph.D. in geology and geophysics from the California Institute of Technology.

William J. Patterson is an independent consultant in areas relating to the design, development, and testing of nuclear and conventional weaponry. He retired from Sandia National Laboratories in 1996 after more than 35 years of service in lead engineering and management roles. Prior to his retirement, he served as manager of the B61 tactical and strategic bomb stockpile and as special projects manager. In the latter capacity he led conventional and nuclear earth-penetrating weapons design efforts. Mr. Patterson received a B.S. and an M.S. in mechanical engineering from Auburn University and the University of New Mexico, respectively.

Gloria S. Patton is currently an independent consultant whose research interests include chemical weapons demilitarization, counterterrorism, and counterproliferation, including areas relating to nuclear, chemical, and biological warfare. From 1999 to 2001, she served as Deputy Assistant Secretary of the Army for Chemical Demilitarization, having served earlier as Principal Deputy (1998 to 1999) and as senior policy advisor to the Assistant to the Secretary of Defense for Nuclear, Chemical and Biological Defense and the Assistant Secretary of Energy for National Security and Nonproliferation (1996 to 1998). In addition, she served as Associate Deputy Assistant Secretary of Technology Development for Environmental Management at the Department of Energy (1990 to 1993), as well as program manager of the Navy Chemical, Biological, and Radiological Defense Advanced and Full Scale Engineering Development Program (during the 1980s). Dr. Patton is a member of the Endocrine Society, Mayo Clinic Alumni, American Physiological Society, Sigma Xi, the Senior Executive Association, and many other professional organizations. She received an M.S. in biochemistry and a Ph.D. in physiology from the University of Southern California.

Heinz W. Schmitt is an independent consultant in areas relating to weapon systems, both of conventional and nuclear origin. He retired from Sandia National Laboratories in 1998, where his last position was Vice President, Weapons Systems. Dr. Schmitt's career at Sandia spanned a variety of technical interests related to weapons systems, to include quality control, computing, component design, structural and dynamic analysis, and systems evaluation. In 1991 he served as vice president for component development and engineering support, which included oversight in design definition, subsystems development, testing, intelligent systems, robotics, and manufacturing. He is fellow of ASME and a recipient of the Secretary of Defense Medal for Outstanding Public Service. Dr. Schmitt received degrees in mechanical engineering, including a B.S. from Brooklyn Polytechnic Institute, an M.S. from the University of New Mexico, and a Ph.D. from Oklahoma State University.

Eugene Sevin is an independent consultant in areas relating to nuclear and conventional weaponry effects, hardened facility design, and computational structural mechanics. He formerly served with the U.S. Department of Defense (DOD) as Deputy Director, Space and Missiles Systems and with the Defense Nuclear Agency as Assistant to the Deputy Director (Science and Technology) for Experimental Research. Prior to joining the DOD, he was a professor of mechanical engineering at the Technion, Israel Institute of Technology; head, mechanical engineering at Ben Gurion University of the Negev, Israel; adjunct professor of applied mechanics at Illinois Institute of Technology (IIT); and director of engineering mechanics research at IIT's Research Institute. Dr. Sevin has served on numerous scientific boards and advisory committees, including as chair of the committee that produced the 1995 NRC report *Protecting Buildings from Bomb Damage*. More recently, he served on a peer review group for the U.S. Army Corps of Engineers, Waterways Experiment Station, and a Defense Science Board task force on underground facilities. He is a member of the National Academy of Engineering. Dr. Sevin received a B.S. in mechanical engineering from IIT, an M.S. in mechanical engineering from the California Institute of Technology, and a Ph.D. in applied mechanics from IIT.

C. Bruce Tarter is director emeritus of Lawrence Livermore National Laboratory, University of California, where his research interests include risk assessment, statistical mechanics, atomic physics, high-temperature equations of state and opacities, theoretical astrophysics, and applications to fusion and defense problems. He has served on numerous scientific boards and advisory committees, including the Laboratory Operations Board (Secretary of Energy Advisory Board), the Nuclear Energy Research Advisory Committee, and the Council on Foreign Relations. He received an S.B. in physics from the Massachusetts Institute of Technology and a Ph.D. in theoretical physics from Cornell University.

Robert H. Wertheim is a retired Rear Admiral, U.S. Navy. During his 35-year career in the Navy, Admiral Wertheim served as director of Navy Strategic Systems projects, responsible for the research, development, production, and operational support of the Navy's submarine-launched ballistic missile systems—Polaris, Poseidon, and Trident. After retirement from the Navy in 1980, he spent 7 years as senior vice president of science and engineering at Lockheed Corporation, and since then has been an independent consultant. Admiral Wertheim has served on numerous scientific boards and advisory committees, including as a member of the NAS/NRC Committee on International Security and Arms Control (from 1989 to 1997). Today, he serves as a member of the Strategic Advisory Group of the U.S. Strategic Command, Joint Department of Defense/Department of Energy Advisory Committee on Nuclear Weapons Surety, and the University of California President's National Security Panel. He is a member of the National Academy of Engineering. Admiral Wertheim received a B.S. from the U.S. Naval Academy and an M.S. from the Massachusetts Institute of Technology.

Staff

James E. Killian is a senior program officer at the National Research Council's National Materials Advisory Board and a retired U.S. Navy captain. During his 26-year career in the Navy he served as the commanding officer of an aircraft carrier-based A-7 Corsair II squadron, commanding officer of the Navy's Nuclear Weapons Evaluation Facility in Albuquerque, New Mexico, and program manager for the Navy's Theater Nuclear Warfare Program (PMS-423) in Washington, D.C. He has a B.S. from the U.S. Naval Academy in Annapolis, Maryland, and an M.S. in aeronautical engineering from the Naval Postgraduate School in Monterey, California.

B

Agendas

KECK CENTER OF THE NATIONAL ACADEMIES
WASHINGTON, D.C.

Tuesday, February 17, 2004

Closed Session: Committee Members and NRC Staff Only

0830 CONVENE—Welcome, Opening Remarks, Introductions
John Ahearne, Committee Chair
Peter Blair, Executive Director, Division on Engineering and Physical Sciences (DEPS)
Charles Draper, Acting Director, Naval Studies Board (NSB)
James Killian, Study Director

0900 COMPOSITION AND BALANCE DISCUSSION
Dennis Chamot, Associate Executive Director, DEPS

Data-Gathering Meeting Not Open to the Public: Classified Discussion

1045 INTRODUCTION OF ATTENDEES AND AGENDA OUTLINE
John Ahearne, Committee Chair
Milton Minneman, Office of the Under Secretary of Defense for Acquisition, Technology, and Logistics (USD (AT&L))/Defense Systems
LTC Jeffrey Davis, USA, Department of Energy (DOE), National Nuclear Security Administration (NNSA)

1100 DOD VIEWPOINT—The New Triad, Hard and Deeply Buried Targets, Key Performance Parameters, Summer 2002 Assessment, Need for Nuclear Earth-Penetrating Weapons
Gregory Hulcher, Special Assistant to the Director, USD (AT&L)/Defense Systems
MAJ Mark Wittig, USA, Nuclear Weapons Employment Analyst, Advanced Weapons Concept Team, Strategic Studies and Analysis Branch, U.S. Strategic Command (USSTRATCOM)

1230 DOE VIEWPOINT—Stockpile Maintenance, Robust Nuclear Earth Penetrator (RNEP) Relationship
David Crandall, Assistant Deputy Administrator for Research, Development, and Simulation, DOE, NNSA

1300 TARGET DEFEAT ASSESSMENT
Milton Minneman, USD (AT&L)/Defense Systems
MAJ Mark Wittig, USA, Nuclear Weapons Employment Analyst, Advanced Weapons Concept Team, Strategic Studies and Analysis Branch, USSTRATCOM
Michael Giltrud, Defense Threat Reduction Agency (DTRA)

1400 ROBUST NUCLEAR EARTH PENETRATOR WEAPON KEY PERFORMANCE PARAMETERS AND DESIGN FEASIBILITY STUDY
Frank Fairchild, U.S. Air Force Nuclear Weapons and Counterproliferation Agency
R. Glenn Bell, Office of Stockpile Assessment and Certification, DOE

1515 TARGET VULNERABILITY TO CONVENTIONAL WEAPONS
Michael Giltrud, DTRA

1600 ADDITIONAL QUESTION AND ANSWER PERIOD
Moderator: John Ahearne, Committee Chair

Closed Session: Committee Members and NRC Staff Only

1630 COMMITTEE DISCUSSION—Day One Summary and Impressions
Moderator: John Ahearne, Committee Chair

1730 END SESSION

Wednesday, February 18, 2004

Closed Session: Committee Members and NRC Staff Only

0830 CONVENE—Opening Remarks, Committee Discussion, Study Plans, Report Deliberations
Moderator: John Ahearne, Committee Chair

Data-Gathering Meeting Not Open to the Public: Classified Discussion

0845 TARGET VULNERABILITY TO NUCLEAR WEAPONS
Michael Giltrud, DTRA

0930 FALLOUT CALCULATION METHODOLOGY
Donald Linger, DTRA

1030 WEAPON OUTPUT AND ENVIRONMENTS
Thomas Thomson, Lawrence Livermore National Laboratory

Closed Session: Committee Members and NRC Staff Only

1115 COMMITTEE DISCUSSION—Continue Committee Discussion, Study Plans, Report Deliberations
Moderator: John Ahearne, Committee Chair

1500 ADJOURN

SANDIA NATIONAL LABORATORIES
ALBUQUERQUE, NEW MEXICO

Tuesday, March 23, 2004

Data-Gathering Meeting Not Open to the Public: Classified Discussion (Secret)

0800 WELCOME, LABORATORY OVERVIEW
John H. Stichman, Vice President, Weapons Systems Division, Sandia National Laboratories

0815 EFFECTS OF FALLOUT AND OTHER RADIOACTIVE MATERIAL
Fred A. Mettler, M.D., New Mexico Veterans Administration Health Care System

0915 CONVENTIONAL HIGH EXPLOSIVES EARTH-PENETRATOR WEAPONS
C. Wayne Young, Applied Research Associates, Inc.

1000 LONG-TERM EFFECTS OF FALLOUT
Dana Powers, Sandia National Laboratories

1100 B-61 MODIFICATION REQUIREMENTS AND TESTING
Kevin Eklund, Distinguished Member, Technical Staff, Sandia National Laboratories

1300 NUCLEAR EARTH-PENETRATOR WEAPON EFFECTIVENESS AND EFFECTS—Joint Laboratory Perspective
Jon Rogers, Manager, National Security Studies Department, Sandia National Laboratories
Rodman R. Linn, Los Alamos National Laboratory
Ted F. Harvey, Lawrence Livermore National Laboratory

1445 B-61/B-83 RNEP BRIEFING
Alfredo McDonald, Manager, B-83 and Penetrator Systems Engineering, Lawrence Livermore National Laboratory

Closed Session: Committee Members and NRC Staff Only

1630 IMPRESSIONS FROM DAY ONE
Moderator: John Ahearne, Committee Chair

DEFENSE THREAT REDUCTION AGENCY
KIRTLAND AIR FORCE BASE, NEW MEXICO

Wednesday, March 24, 2004

Data-Gathering Meeting Not Open to the Public: Classified Discussion

0800 AGENT DEFEAT
Donald Linger, Deputy for Test and Technology, DTRA

0850 OPERATION IRAQI FREEDOM GROUND TRUTH
Thomas Lutton, Senior Scientist, DTRA
Robert Delisle, Contractor, DTRA

1000 DEFENSE THREAT REDUCTION AGENCY OPERATIONAL HAZARD PREDICTION CODE—Hazard Prediction and Assessment Capability (HPAC)
Todd Hann, Deputy for Consequence Assessment Branch, DTRA
1100 SUBSTANTIAL SHAPED CHARGE PROCEEDING NUCLEAR EARTH-PENETRATOR WEAPON
Thomas Togami, Advanced and Exploratory Systems Department, Sandia National Laboratories

Closed Session: Committee Members and NRC Staff Only

1230 COMPOSITION AND BALANCE DISCUSSION—Update to Meeting One
1300 COMMITTEE DELIBERATIONS; SUMMARY DISCUSSION
Moderator: John Ahearne, Committee Chair
1500 ADJOURN

KECK CENTER OF THE NATIONAL ACADEMIES WASHINGTON, D.C.

Tuesday, April 27, 2004

Closed Session: Committee Members and NRC Staff Only

0755 CONVENE—Welcome, Opening Remarks, Introductions
John Ahearne, Committee Chair
James Killian, Study Director

Data-Gathering Session Not Open to the Public: Classified Discussion

0800 SECOND-HAND THOUGHTS ON TARGETING BIOLOGICAL AGENTS
David R. Franz, Chief Biological Scientist, Midwest Research Institute

Data-Gathering Session Open to the Public: Unclassified Discussion

0930 EARTH-PENETRATING NUCLEAR WEAPONS—TARGET DAMAGE AND FALLOUT
Thomas B. Cochran, Director, Nuclear Programs, Natural Resources Defense Council
1000 EFFECTIVENESS OF EARTH-PENETRATING WEAPONS FOR "BUNKER BUSTING" AND AGENT DEFEAT
Robert W. Nelson, Senior Fellow, Science and Technology, Council on Foreign Relations
1030 AGENT DEFEAT CALCULATIONS
Michael A. Levi, Science and Technology Fellow in Foreign Policy Studies, Brookings Institution
1115 CONSEQUENCES OF THE JAPANESE ATOMIC BOMBINGS
Mortimer L. Mendelsohn, Senior Scientist, Biology and Biotechnology Research Program, Lawrence Livermore National Laboratory

Data-Gathering Session Not Open to the Public: Classified Discussion

1300 CONVENTIONAL SOURCES—Target Release and Agent Defeat
Todd H. Pierce, Science Applications International Corporation

1345 UNCERTAINTIES IN CHEMICAL AND BIOLOGICAL SOURCE TERM MODELS FOR NUCLEAR ATTACKS ON WEAPONS OF MASS DESTRUCTION FACILITIES
Philip A. Hookham, Titan Corporation

1445 FIRST PRINCIPLES MODELING OF CLOUD RISE AND DISPERSION
David P. Bacon, Science Applications International Corporation
R. Ian Sykes, Titan Corporation

1615 FALLOUT PHENOMENOLOGY IN HPAC
Joseph T. McGahan, Science Applications International Corporation

Closed Session: Committee Members and NRC Staff Only

1715 COMMITTEE DELIBERATIONS—Day One Summary and Impressions
Moderator: John Ahearne, Committee Chair

1730 END SESSION

Wednesday, April 28, 2004

Closed Session: Committee Members and NRC Staff Only

0755 CONVENE—Welcome, Opening Remarks, Introductions
John Ahearne, Committee Chair
James Killian, Study Director

Data-Gathering Session Open to the Public: Unclassified Discussion

0800 MEETING WITH CONGRESSIONAL STAFF

Data-Gathering Session Not Open to the Public: Classified Discussion

0915 AGENT DEFEAT CAPABILITY AND ASSOCIATED COLLATERAL DAMAGE OF CONVENTIONAL AND NUCLEAR WARHEADS
Hans W. Kruger, Lawrence Livermore National Laboratory

1015 AGENT DEFEAT WEAPON FINDINGS/LESSONS LEARNED
Michael A. Martinez, Chief, Advanced Technology Division, Air Force Nuclear Weapons and Counterproliferation Agency

1130 RESULTS OF HAZARD PREDICTION ANALYSIS CODE SAMPLE CALCULATIONS
Michael Phillips, Contractor, DTRA

1300 RESULTS OF K DIVISION FALLOUT CODE/LAGRANGIAN OPERATIONAL DISPERSION INTEGRATOR (KDFOC/LODI) SAMPLE CALCULATIONS
Ted F. Harvey, Lawrence Livermore National Laboratory

Closed Session: Committee Members and NRC Staff Only

1415 COMMITTEE DELIBERATIONS—Study Plans, Report Deliberations
Moderator: John Ahearne, Committee Chair
1600 ADJOURN

KECK CENTER OF THE NATIONAL ACADEMIES
WASHINGTON, D.C.

Monday, May 17, 2004

Data-Gathering Session Not Open to the Public: Classified Discussion

0830 CONVENE—Welcome, Opening Remarks, Introductions
John Ahearne, Committee Chair
James Killian, Study Director
0835 THE NEED FOR RNEP
Gregory Hulcher, Special Assistant to the Director, USD (AT&L))/Defense Systems
0905 EFFECTS OF NUCLEAR EARTH-PENETRATOR WEAPONS
Bryan Fearey, Los Alamos National Laboratory
John St. Ledger, Los Alamos National Laboratory

Data-Gathering Session Open to the Public: Unclassified Discussion

1000 CONGRESSIONAL RESEARCH SERVICE PERSPECTIVE
Jonathan Medalia, Congressional Research Service
1045 HEALTH AND ENVIRONMENTAL EFFECTS OF BIOLOGICAL WEAPONS
Stephen Morse, Associate Director for Science, Bio-Terrorism, Preparedness and Response Program, Centers for Disease Control and Prevention
1130 RNEP
Keith Payne, National Institute for Public Policy

Data-Gathering Session Not Open to the Public: Classified Discussion

1300 COLLATERAL RISKS FROM ATTACKS ON CHEMICAL AND BIOLOGICAL FACILITIES
Jeffrey Grotte, Deputy Director, Strategy, Forces, and Resources Division, Institute for Defense Analyses
1345 BATTLEFIELD OPERATIONS IN A NUCLEAR, BIOLOGICAL, AND CHEMICAL (NBC) ENVIRONMENT
COL Richard Hill, USA, Operations Division Chief, Nuclear Division, U.S. Army Nuclear and Chemical Agency
Martin Moakler, Physical Scientist, U.S. Army Nuclear and Chemical Agency
1445 HARD TARGET DEFEAT
Donald Linger, Deputy for Test and Technology, DTRA
1530 HPAC/ANRAC COMPUTER RUNS DISCUSSION
Todd Hann, DTRA

1630 CONVENTIONAL PENETRATORS
Barry Hannah, Branch Head Reentry Systems, Navy Strategic Systems Programs

Closed Session: Committee Members and NRC Staff Only

1700 COMMITTEE DELIBERATIONS—Day One Summary and Impressions, Outline and Writing Assignments Status
Moderator: John Ahearne, Committee Chair
1900 COMMITTEE DELIBERATIONS—Discuss Preliminary Findings and Recommendations
Moderator: John Ahearne, Committee Chair
2100 END SESSION

Tuesday, May 18, 2004

Data-Gathering Session Not Open to the Public: Classified Discussion

0800 CONVENE—Welcome, Opening Remarks, Introductions
John Ahearne, Committee Chair
James Killian, Study Director
0805 HEALTH EFFECTS OF CHEMICAL WARFARE AGENTS
David Moore, Vice President of Medical Technology, Battelle Memorial Institute
0905 BIOLOGICAL WEAPONS AEROSOLIZED—Viable vs. Respirable
Martin Bagley, DTRA

Closed Session: Committee Members and NRC Staff Only

1000 COMMITTEE DELIBERATIONS—Preparation for Woods Hole Meeting
John Ahearne, Committee Chair
James Killian, Study Director
1200 ADJOURN

**J. ERIK JONSSON WOODS HOLE CENTER
WOODS HOLE, MASSACHUSETTS**

Monday, June 21, 2004

Data-Gathering Session Not Open to the Public: Classified Discussion

0830 CONVENE—Welcome, Opening Remarks, Introductions
John Ahearne, Committee Chair
James Killian, Study Director
0835 NUCLEAR EVENT CALCULATIONS
Todd Hann, DTRA

Closed Session: Committee Members and NRC Staff Only

1300 REPORT DISCUSSION AND DRAFTING
Moderator: John Ahearne, Committee Chair
1700 END SESSION

Tuesday Through Thursday, June 22-24, 2004

Closed Session: Committee Members and NRC Staff Only

0830 CONVENE—Plans for the Day
John Ahearne, Committee Chair
James Killian, Study Director
0835 REPORT DISCUSSION AND DRAFTING
Moderator: John Ahearne, Committee Chair
1300 REPORT DISCUSSION AND DRAFTING (CONTINUED)
Moderator: John Ahearne, Committee Chair
1700 END SESSION

C

Equivalent Yield Factors for Energy Coupling

The energy coupled to the ground from a near-surface nuclear burst is plotted in Figure 4.1 as an equivalent yield factor. The functional relationships for these results are based on a set of algorithms for energy coupling endorsed by the Defense Threat Reduction Agency that account for the effects of weapon design, geologic media, and source location[1] and are summarized below.

Equivalent Yield Based on Coupled Energy

$$f_e(d_s) := \begin{cases} 0.0146\cdot\left(\left|d_s\right|\right)^{-0.106} - 0.01 & \text{if } d_s < -0.05 \\ \text{otherwise} & \\ \quad \begin{cases} \min\left(0.504\cdot d_s^{0.36} - 0.01, 1\right) & \text{if } d_s > 0.05 \\ 0.059\cdot e^{21.455\cdot d_s} - 0.01 & \text{otherwise} \end{cases} \end{cases}$$

Equivalent Yield Based on Ground Shock (with air-blast effects)

$$f_{gs}(d_s) := \begin{cases} 0.0146\cdot\left(\left|d_s\right|\right)^{-0.106} & \text{if } d_s < -0.05 \\ \text{otherwise} & \\ \quad \begin{cases} \min\left(0.504\cdot d_s^{0.36}, 1\right) & \text{if } d_s > 0.05 \\ 0.059\cdot e^{21.455\cdot d_s} & \text{otherwise} \end{cases} \end{cases}$$

where

f_e = equivalent yield coupled energy

f_{gs} = equivalent ground-shock-coupled energy

d_s = the scaled depth of burst ($m/kt^{1/3}$), positive into ground.

These relationships are based on numerical calculations as the only source of energy-coupling data.

NOTE

1. Defense Nuclear Agency. 1991. *Effects Manual Number 1,* Chapter 3, "Cratering, Ejecta, and Ground Shock," DNA-EM-1-CH-3, Alexandria, Va., December.

D

Acronyms and Abbreviations

ACTD	Advanced Concept Technology Demonstration
AF&F	Arming, Fusing, and Firing
ATACMS	Army Tactical Missile System
ATD	atmospheric transport and dispersion
AUP	Advanced Unitary Penetrator
BW	biological weapon
C^3	command, control, and communications
C^3I	command, control, communications, and intelligence
CBRN	chemical, biological, radiological, and nuclear
CEP	circular error probable
CRH	caliber radius head
CW	chemical weapon
DELFIC	Defense Land Fallout Interpretive Code
DIA	Defense Intelligence Agency
DOB	depth of burst
DOD	Department of Defense
DOE	Department of Energy
DTED	digital topographic elevation data
DTRA	Defense Threat Reduction Agency
DUG1c	one-dimensional engineering ground-shock propagation code
EM-1	*Effects Manual-1*
EP	earth penetration
EPW	earth-penetrator weapon

GPS	Global Positioning System
GVN	ground vulnerability number
HDBT	hard and deeply buried target
HOB	height of burst
HPAC	Hazard Prediction and Assessment Capability (code)
ICRP	International Commission on Radiological Protection
JDAM	Joint Direct Attack Munition
KDFOC	K-Division Defense Nuclear Agency Fallout Code
LANL	Los Alamos National Laboratory
LD^{50}	level of radiation at which 50 percent of the population would die
LLNL	Lawrence Livermore National Laboratory
NCRP	U.S. National Council on Radiation Protection and Measurements
NEWFALL	New Fallout Code
NGA	National Geospatial-Intelligence Agency
NNSA	National Nuclear Security Administration
NTS	Nevada Test Site
OMEGA	Operational Multiscale Environment Model with Grid Adaptivity
P II	Pershing II (missile)
PD	probability of damage
PDCALC	Probability of Damage Calculator
R&D	research and development
RIPD	Radiation Induced Performance Distribution
RNEP	robust nuclear earth penetrator (weapon)
SCIPUFF	Second-Order Closure Integrated Puff
SDOB	scaled depth of burial
SEPW	strategic earth-penetrator weapon
SNL	Sandia National Laboratories
SNLA	Sandia National Laboratories Albuquerque
SNLL	Sandia National Laboratories Livermore
SRBM	short-range ballistic missile
SWIFT	Stationary Wind Fit and Turbulence
UGT	underground nuclear test
USSTRATCOM	United States Strategic Command
UV	ultraviolet

VN	vulnerability number
WinGS	two-dimensional physics-based ground-shock code
WMD	weapons of mass destruction